The Objects of Credence

The Objects of Credence

ANNA MAHTANI

UNIVERSITY PRESS

Great Clarendon Street, Oxford, OX2 6DP,
United Kingdom

Oxford University Press is a department of the University of Oxford.
It furthers the University's objective of excellence in research, scholarship,
and education by publishing worldwide. Oxford is a registered trade mark of
Oxford University Press in the UK and in certain other countries

© Anna Mahtani 2024

The moral rights of the author have been asserted

All rights reserved. No part of this publication may be reproduced, stored in
a retrieval system, or transmitted, in any form or by any means, without the
prior permission in writing of Oxford University Press, or as expressly permitted
by law, by licence or under terms agreed with the appropriate reprographics
rights organization. Enquiries concerning reproduction outside the scope of the
above should be sent to the Rights Department, Oxford University Press, at the
address above

You must not circulate this work in any other form
and you must impose this same condition on any acquirer

Published in the United States of America by Oxford University Press
198 Madison Avenue, New York, NY 10016, United States of America

British Library Cataloguing in Publication Data

Data available

Library of Congress Control Number: 2023945528

ISBN 978–0–19–884789–2

DOI: 10.1093/oso/9780198847892.001.0001

Printed and bound in the UK by
Clays Ltd, Elcograf S.p.A.

Links to third party websites are provided by Oxford in good faith and
for information only. Oxford disclaims any responsibility for the materials
contained in any third party website referenced in this work.

Contents

Acknowledgements	ix
1. Introduction	1
2. Propositions	7
2.1 Introduction	7
2.2 The roles that propositions play	7
2.2.1 Propositions are the objects of propositional attitudes	7
2.2.2 Propositions are the contents of declarative utterances	10
2.2.3 Propositions are truth-apt	12
2.3 Frege's account	13
2.4 Russell's account	18
2.5 Guise Russellianism	20
2.6 Possible world semantics	22
2.7 Chapter summary	29
3. The Credence Framework	31
3.1 Introduction	31
3.2 The probability framework	32
3.3 What are credences?	35
3.4 The credence framework	37
3.5 Arguments for the probability axioms	40
3.6 Other rules of rationality	44
3.7 Chapter summary	47
4. Credence Claims Are Opaque	49
4.1 Introduction	49
4.2 The 'it's just obvious' argument	49
4.3 Choice behaviour	52
4.4 Omniscience	57
4.5 Conditionalization	60
4.6 Guise Russellianism	62
4.7 A guise-based account of credence	66
4.8 A different guise-based account of credence	69
4.9 Chapter summary	72

vi CONTENTS

5. Implications for Rationality 73
  5.1 Introduction 73
  5.2 The Reflection Principle 74
    5.2.1 The Reflection Principle and the Generalized
        Reflection Principle 75
    5.2.2 The Mug 77
    5.2.3 The Improved Generalized Reflection Principle 79
    5.2.4 Sleeping Beauty 84
    5.2.5 Other principles of deference and disagreement 86
  5.3 The Principal Principle 87
    5.3.1 Chance claims are not extensional 88
    5.3.2 The chance framework 92
    5.3.3 The Principal Principle and the contingent a priori 94
    5.3.4 Old problems for the Principal Principle 97
    5.3.5 The New New Principle 101
  5.4 Chapter summary 104

6. Practical Implications 105
  6.1 Introduction 105
  6.2 Decision theory 105
  6.3 Decision theory and the objects of credence 108
  6.4 The two-envelope paradox 111
    6.4.1 A specific version of the two-envelope paradox 113
    6.4.2 'M' as a transparent designator 114
    6.4.3 'M' as a definite description 115
    6.4.4 Are rigid designators safe to use as outcomes in
        decision tables? 117
    6.4.5 A better restriction 118
    6.4.6 Variations 123
  6.5 Welfare economics 125
    6.5.1 Utilitarianism, egalitarianism, and prioritarianism 127
    6.5.2 The Pareto principle 128
    6.5.3 Prospects for individuals and opacity 131
    6.5.4 Supervaluationism 135
  6.6 Chapter summary 136

7. States as Metaphysically Possible Worlds 138
  7.1 Introduction 138
  7.2 Russell's descriptivism 140
  7.3 Stalnaker's account of belief attribution 142
  7.4 Chalmers's account—the basics 146
  7.5 The objects of credence as primary intensions 152
  7.6 A new convention for credence attribution statements 157

CONTENTS vii

7.7	The objects of credence as primary intensions and secondary intensions	162
7.8	The objects of credence as enriched propositions	166
7.9	Chapter summary	168

8. States as Something Else 169
   8.1 Introduction 169
   8.2 Fine-grained worlds 170
   8.3 Linguistic representations 175
   8.4 Sentence-worlds 177
   8.5 Sentence-context-worlds 179
   8.6 A world-building language 181
   8.7 Completeness and coherence 184
   8.8 A stronger notion of coherence 188
   8.9 Arguments for probabilism 191
   8.10 Structured propositions 193
   8.11 Chapter summary 196

9. Conclusion 197

References 199
Index 207

Acknowledgements

I offer very grateful thanks to all the people who commented on ideas or drafts. In particular (and with apologies to anyone I may have overlooked): Matthew D. Adler, Arif Ahmed, Marius Backmann, Nicholas Baigent, Mike Beaney, Jonathan Birch, Jens Christian Bjerring, Luc Bovens, Richard Bradley, David Braun, Liam Kofi Bright, Campbell Brown, Heather Browning, Susanne Burri, Nicholas Côté, David Chalmers, Chloé De Canson, Edward Elliott, Melissa Fusco, Giacomo Giannini, Ze'ev Goldschmidt, Alan Hájek, Caspar Hare, Anandi Hattiangadi, Terry Horgan, Laurenz Hudetz, Mark Jago, Rosanna Keefe, Sophie Kikkert, Ko-Hung Kuan, Maria Lasonen-Aarnio, Christian List, Ofra Magidor, Nicholas Makins, Silvia Milano, Richard Pettigrew, Andrea Petrou, Miklos Redei, David Ripley, Bryan Roberts, Lewis Ross, Daniel Rothschild, Joe Roussos, Nathan Salmon, Moritz Schulz, Katie Steele, Scott Sturgeon, Johanna Thoma, Somayeh Tohidi, Aron Vallinder, Alex Voorhoeve, Kate Vredenburgh, Robbie Williams, and Timothy Williamson.

I'd also like to thank the Leverhulme Trust for funding me while I worked on many of these ideas, and finally also Ben, Sam, and Kate for being wonderful interlocutors!

1

Introduction

We all experience uncertainty. We don't know how things will be in ten years' time, and we can't even be certain what tomorrow will bring. How can we make decisions in this state of uncertainty? And how can we predict and explain what other people will do, given that they are uncertain too? The near-universal response is to rely on the credence framework.

To introduce the credence framework, I begin by contrasting credences with beliefs. To believe a proposition is an all or nothing matter: either you believe it or you don't.[1] The credence framework introduces gradation. You might have a credence of 1 in a proposition (if you are completely certain of it) or a credence of 0 in a proposition (if you are completely certain that it is *not* the case), or any number in between.[2] Perhaps you are uncertain whether tomorrow's train will be cancelled—let's say you have a credence of 0.8 that it will be cancelled, and a credence of 0.2 that it won't be. This can help us predict and explain your behaviour: for example, we may be able to use your credences to predict that you won't bother going to the station and will decide to work from home instead; and we can point to your credences to explain why you did so. And thinking about your credences may also help you work out what to do: you may figure out that—given your credences—it would be rational to stay at home.

In my interactions and research at the LSE I encounter this framework almost every day. This may be extreme (the LSE department of Philosophy, Logic, and Scientific Method is admittedly an unusual place!) but the framework is enormously prevalent elsewhere too. It is used by scientists and social scientists in almost all disciplines, including economics and political theory, and it underpins policy choice in healthcare, transport, education, and numerous other areas. It is hard to overestimate its importance. There are subtle and important differences in the way that

---

[1] This is not to deny that 'believes' might be a vague predicate: my point is just that it does not explicitly introduce a scale.

[2] Some distinguish between having a credence of 1 in some claim and being certain of that claim (and similarly between having a credence of 0 in some claim and being certain that it is not the case). Here I talk about certainty for simplicity.

The Objects of Credence. Anna Mahtani, Oxford University Press. © Anna Mahtani 2024.
DOI: 10.1093/oso/9780198847892.003.0001

2 INTRODUCTION

these different disciplines use the idea of credences, and there is some different terminology too—including 'degrees of belief', 'subjective probabilities', and 'Bayesian epistemology'—but the underlying idea is roughly the same: whenever a person has limited evidence, and so faces uncertainty, we represent her epistemic state by assigning numbers to propositions to represent her credence in each.

Before I was immersed in a world where this framework was the norm, I used to work in the philosophy of language. It might seem that there are unlikely to be any important links between the philosophy of language and the credence framework, but soon after I encountered the framework I was struck by what seemed an enormously important question: what are the objects of credence? That is, when we give the numbers that represent a person's credences, what are we assigning these numbers to? My focus in this book is on this question. Above I stated that people have credences in propositions, but what exactly are propositions? Theorists working in the philosophy of language have been vigorously debating this question for many years, and there is widespread and unresolved disagreement.[3] Some argue that propositions are sets of possible worlds; some argue that they cannot be sets of possible worlds, but must be sets of something different—such as impossible worlds, or sentences; others argue that propositions cannot be sets of worlds at all, but are rather structured entities; and others maintain that there are no such things as propositions. Those using the credence framework, in contrast, have not generally focused on this question. Many are content with a formal answer which follows from the details of the framework: a proposition is an 'event', which is a set of 'states'. But the problems and complexities that are discussed by philosophers of language are all applicable here, and while this battle rages over the nature of propositions, we cannot be complacent about the credence framework. The objects of credence are presumably the same sorts of things as the objects of belief—namely propositions. If propositions are sets of states, as those using the credence framework claim, then what are these states? And if propositions are *not* sets at all, then what will happen to the credence framework? And—even more radically—if there are no such things as propositions, as some philosophers of language claim, what then? These are open questions.

One concrete issue that is of particular importance for this book is whether credence claims are transparent or opaque. An illustration will help to explain this:

---

[3] For an overview of the current state of the debate, see (King, Soames, and Speaks, 2014).

(1a)   Tom has a credence of 0.8 that George Orwell is a writer.

(1b)   Tom has a credence of 0.1 that Eric Blair is a writer.

Sentences (1a) and (1b) above are both 'credence attribution' statements: they attribute to Tom particular credences in particular propositions. All those who use the credence framework make attribution statements like this. We use them to predict and explain peoples' behaviour; and we use them (often in the first person) to decide what to do. The key question for us here is: is it possible for both (1a) and (1b) to be true together (with 'Tom' picking out the same person in each sentence, and the relevant time being the same, and so on)? That is, can Tom have a high credence that George Orwell is a writer, while simultaneously having a low credence that Eric Blair is a writer? George Orwell and Eric Blair are actually one and the same: George Orwell was born 'Eric Blair', and took the name 'George Orwell' as a pen name. This might make you think that if (1a) is true, then (1b) can't be—for how can Tom's credence that George Orwell is a writer be different from his credence that Eric Blair is a writer, given that George Orwell *is* Eric Blair? On this view, a person's credences are about *things in themselves*. Thus Tom's credence of 0.8 that George Orwell is a writer is a credence *about that person*, and we could express Tom's epistemic state equally well by saying that he has a credence of 0.8 that Eric Blair is a writer. On this view it doesn't make any difference whether we use the name 'George Orwell' or 'Eric Blair' in our credence attribution statements—for both are just ways to pick out the same person. We can say that on this view credence claims are transparent: the names and so on are just ways of reaching out to the relevant objects. In contrast, you might think—as I do—that both (1a) and (1b) can be true together. Tom might be familiar with the works of George Orwell, and have heard the name 'Eric Blair' in passing, without realizing that Eric Blair and George Orwell are one and the same—in which case he might be pretty sure that George Orwell is a writer while thinking (in the absence of any reason to think otherwise) that Eric Blair is unlikely to be a writer. On this view, credence claims are opaque: when we say what or whom someone's credence is about, it can matter what name or other designator we use. I call this claim—that credence claims are opaque—the 'tenet', and in this book I argue for it, trace its implications, and set out the foundations of the credence framework in the light of it.

The first part of the book is introductory. After this short general overview (chapter 1), I spend the next two chapters introducing in more depth

4 INTRODUCTION

the two very different perspectives that this book brings together. First (in chapter 2) I outline the debate in the philosophy of language over the nature of propositions. Then (in chapter 3) I describe the credence framework and briefly mention some of the uses to which it is put.

The second part of the book is a single chapter (4), in which I argue for the tenet that credence claims are opaque. You might think (as I do) that the tenet is obviously true, but the implications of accepting it are wide-reaching and surprising, and this may give you pause, so it is important that I shore up your acceptance of the claim by providing compelling arguments, and solid reasons to reject the alternative.

In the third part of the book I trace some of the implications of the tenet. First (in chapter 5) I trace some of the implications for principles of rationality: in particular, I look at the implications for various deference principles, and for the Principal Principle—principles that I set out and explain in this chapter. Then (in chapter 6) I describe some of the practical implications for decision theory (where for concreteness I focus on a particular problem called the 'Two-Envelope Paradox'), and welfare economics.

Finally, in the fourth part of the book, I turn to the foundations of the credence framework. Is there a viable interpretation of the credence framework that can accommodate the tenet? Or does the framework need to be radically rethought? I begin (in chapter 7) by exploring whether we can accommodate the tenet by interpreting the framework using tools from two-dimensionalist accounts. And finally (in chapter 8) I consider whether we can accommodate the tenet by interpreting the states in the framework in a particular way. Both broad approaches face serious challenges, and there are wide-reaching implications of which a user of the framework should be aware.

You might wonder why I've decided to organize the book in this way. Why have the implications before the foundations? Surely it would make more sense to propose and motivate the tenet, figure out the foundations, and then trace the implications? My reason for organizing things as I have done is to provoke interest in the topic. I've talked about the tenet to a lot of people who use the credence framework. Many of the people who use the credence framework have only passing familiarity with the philosophy of language: this is unsurprising, given that users of the credence framework extend well beyond philosophers to economists, statisticians, and policy makers, and given that the credence framework has been at least partly constructed in disciplines that are quite independent from the philosophy of language. There are thus many users of the framework who have never

INTRODUCTION 5

explicitly considered the tenet. But whenever I have asked them to consider it (by for example showing them the George Orwell example above), in almost every case they have agreed that credence claims are opaque. In my view, this tenet is very important, with the potential to disrupt many of the principles, arguments, and moves that users of the credence framework take for granted. But showing that the tenet has these major implications takes work. For when I present users of the credence framework with the tenet, I have often had a response along the following lines:

> The credence framework itself is very flexible, and it can be interpreted in all sorts of different ways. For example, when we (the proponents of the credence framework) say that there is an underlying set of 'states' (which form part of the credence framework) that leaves open what sorts of things these states might be; and similarly, when we talk about an 'algebra' (another part of the credence framework), there is no prejudgement about what sorts of things make up this algebra. It may well be that credence claims are opaque as you say, but the credence framework would be able to accommodate this fact without difficulty. Thus we can continue to use the credence framework just as we do now, while in parallel a largely unconnected project can be run on how to interpret the framework so as to accommodate the point that credence claims are opaque.

Thus while there may be some intellectual interest in figuring out how to accommodate the tenet, the issue strikes many users of the credence framework as non-urgent, and unlikely to affect the many projects which continue to make use of the framework. I see things differently. On my view, the tenet forces us to reassess many of the uses to which the credence framework is put: there are concepts which need to be rethought, moves which turn out to be invalid, and principles which need to be rejected or transformed. We cannot then continue to use the framework without paying attention to this issue. I argue for this by showing in chapters 5 and 6 a range of cases where the tenet forces a rethink in how the credence framework is used: these include both concepts and principles connected with rationality, and more practical moves made in decision theory and welfare economics. By establishing that the tenet makes a real difference—a difference that users of the framework need to pay attention to—I motivate interest in the final chapters where I turn to foundational issues.

In writing this book I've had two different sorts of readers in mind. Firstly, I imagine the book being read by a user of the credence framework. For this

6 INTRODUCTION

reader, I aim to introduce the relevant ideas from the philosophy of language, show the connection between these ideas and the credence framework, trace some of the implications of these connections for the work that the user of the credence framework might carry out, and describe how we might rethink the credence framework in the light of them. Secondly, I hope that the book will be of equal interest to philosophers of language and similar disciplines. For these philosophers, it should be interesting to see how ideas that they are familiar with have major implications for the credence framework—implications that we can trace through to practical policy choice.

In line with these two possible sorts of readers, I've found that in presenting work from this book I have met with two very different reactions, depending on whether I am presenting it primarily to philosophers of language or to users of the credence framework (and of course there are philosophers at the intersection of these two groups: I would count myself as a member of this intersection). Philosophers of language are generally familiar with the background to the tenet, and often think it is obvious—though they may be unfamiliar with the way the credence framework is used, and so the wide-reaching implications of the tenet may be unexpected. Users of the credence framework have often never considered the tenet before, but are of course familiar with the way that the credence framework is used. These audiences are sometimes unsettled by the changes that the tenet brings to bear on familiar territory, and surprised to see a point from the philosophy of language disturbing what looks like an unrelated discipline. Such differences in reaction are perhaps only to be expected given that the project is interdisciplinary! My aim in writing this book has been to draw a connection between these two very different fields of vision.

2

Propositions

2.1 Introduction

The aim of this chapter is to introduce the relevant background from the philosophy of language. Our particular focus is on propositions, as these may naturally be assumed to be the objects of credence—although we will encounter other candidates for this role throughout this book.

The idea of a proposition is one of the first things that a philosophy student will come across. As the term is used in philosophy, there are three roles that propositions are expected to play: firstly, they are the objects of propositional attitudes; secondly, they are the content of declarative utterances; and thirdly, they are truth-apt—that is, they are the sorts of things (though perhaps not the only things) that can be true or false. I begin by describing each of these roles in more detail, before turning to outline a range of accounts of propositions.

2.2 The roles that propositions play

2.2.1 Propositions are the objects of propositional attitudes

The first role for propositions is to be the objects of propositional attitudes. Propositional attitudes are attitudes towards propositions, and they include *believing that, desiring that, hoping that, fearing that*—and there are many other examples besides. Consider the following statement:

(2a)   Alice believes that Paris has a population of over 2 million.

Standardly, philosophers take this to mean that Alice stands in a certain attitude—the attitude of belief—towards the proposition that Paris has a

The Objects of Credence. Anna Mahtani, Oxford University Press. © Anna Mahtani 2024.
DOI: 10.1093/oso/9780198847892.003.0002

8 PROPOSITIONS

population of over 2 million. That is why belief is called a 'propositional attitude'—because it is an attitude that one can take towards a proposition.[1]

But what are propositions? And why introduce this strange new idea? Why not just say that Alice's attitude is towards the relevant *sentence*? To develop this thought, we can distinguish between sentence types and sentence tokens. A sentence token is a concrete object—a squiggle on a piece of paper, perhaps, or some sound waves that occur at a particular point in time—but a sentence type (if we think that such a thing exists) will be an abstract rather than a concrete object.[2] Two sentence tokens can be instances of the same sentence type: for example, if you say 'grass is green', and I say 'grass is green', then our two sentence tokens are instances of the same sentence type. Now that we have distinguished sentence tokens from sentence types, we're in a position to interrogate the proposed view on which belief is an attitude taken towards a sentence: is the proposal that a belief is an attitude towards a sentence token, or a sentence type? There are problems with both suggestions.

To see one problem with the first suggestion, consider that you might have beliefs that have never been expressed. You can probably think of some such belief now: I am thinking of my belief that giraffes are generally a lot larger than pencil cases—but of course now that I have expressed it, it no longer serves as an example, and you will have to think of another. When you have a belief that has not been expressed, then there may be no suitable sentence token—no squiggle on a page, no series of sound waves—to be the object of your attitude: your attitude can't be an attitude towards a particular sentence token if no such sentence token exists![3] Let's turn then to the second suggestion, according to which belief is an attitude towards a

---

[1] The word 'believes' is used in various different ways. We might say that Alice believes in God, or believes in the healing power of time. But 'God' does not express a proposition, and nor does 'the healing power of time'. Thus to say that Alice believes in these things is not to say that she stands in particular *propositional* attitudes (though we might think that there is a connection here—for example, we might think that if Alice believes in God then it follows that Alice believes that God exists, and plausibly *that God exists* does express a proposition). For simplicity, in this chapter, I will focus just on the sorts of constructions where 'believes' is followed by a 'that' clause, for it is these sorts of constructions that are most plausibly interpreted as describing propositional attitudes.

[2] A sentence token is similar to what is called an 'utterance', but these are not quite the same thing. One reason to think that they are distinct is that the same sentence token can be used to make several different utterances: 'a sign that says "Flying planes can be dangerous" might first have been used at a pilots' school, to warn would-be pilots, and then recycled on a high hill near an airport, to warn would-be kite-flyers' (Perry, 2003, pp. 377–8).

[3] Some theorists hold that there are token sentences to be found *inside your brain*: these are concrete objects analogous to squiggles on a page. There are many variations and complexities

THE ROLES THAT PROPOSITIONS PLAY   9

sentence type, which is an abstract rather than a concrete object. This avoids
the problem that we had with sentence tokens, but sentence types don't seem
to be quite the right sorts of abstract objects for this role, and understanding
why can help us to see the appeal of propositions. Consider the following
belief attribution:

(2b)   Bertie believes that the apple is in the bowl.

On the view that we are considering, we take this to mean that Bertie stands
in the belief relation to the sentence type of which 'the apple is in the bowl' is
an instance. Yet we could equally well have said that Bertie believes that the
apple is within the bowl, or that the bowl has the apple in it, and many other
rephrasings are possible.[4] Why should these all count as harmless rephrasings, if Bertie's attitude is towards a sentence type? After all, 'the apple is in
the bowl', 'the apple is within the bowl', and 'the bowl has the apple in it' are
all instances of different sentence types. The reason must be that Bertie's
attitude is not towards the sentence type of which 'the apple is in the bowl' is
an instance, but towards something else—the *content* of 'the apple is in the
bowl', which is the same as the content of 'the bowl has the apple in it', and
so forth.[5] This content is what we are calling the proposition, and it can be
expressed by many different sentences. Though there is much debate over
exactly what a proposition is, as we shall see, we can think of it for now as a
state of affairs, or a way things could be. Thus 'the apple is in the bowl'
and 'the bowl has the apple in it' describe the same state of affairs, and
so express the same proposition. Here we've arrived at the standard
view, on which belief is a relation between a person and a proposition.
Statement (2b), for example, states that Bertie stands in the belief relation to
a particular proposition—namely the proposition expressed by 'the apple is

---

within this viewpoint. Some theorists who accept this view aim to dispense with propositions
altogether (Field, 2001), while others maintain an important role for propositions as the content
of these token sentences (Fodor, 1987).

[4] Later in the book I argue that the objects of belief and credence are, in a sense to be
explained, 'fine-grained'. In a similar spirit, we might think (in opposition to the main text here)
that Bertie could believe that the apple is in the bowl without believing that the bowl has the
apple in it. I return to this issue in chapter 8 (section 8.8).

[5] In contrast, consider this claim:

Bertie likes the sound of the sentence 'the apple is in the bowl', and is thinking of
using it in his next poem.

Here we really are talking about Bertie's attitude towards a sentence, and we cannot harmlessly
substitute alternative sentences that have the same content.

10 PROPOSITIONS

in the bowl'—which could equally well have been expressed by a range of other sentences.

Thus tokens that are instances of a range of different sentence types can be used to express the same proposition. The reverse also holds: sentence tokens of the same type can express different propositions. This is because the proposition expressed by a sentence token depends both on the sentence type and on the *context of utterance*. The context will include such factors as the time and place of utterance, the person doing the uttering, and the intentions of the utterer. Suppose, for example, that Bertie says 'the apple is in the bowl'—with a particular apple in mind, and indicating a particular bowl. And now suppose that Bobbie at some other location also says 'the apple is in the bowl'—with a different apple in mind, and indicating a different bowl. Then Bertie and Bobbie have expressed different propositions: Bertie is saying that a particular apple is in a particular bowl, and Bobbie is saying that some other apple is in some other bowl. They are using the same sentence type to express different propositions. Propositions, then, are distinct from both sentence types and sentence tokens. And standardly, the objects of belief are taken to be neither sentence tokens nor sentence types, but rather propositions.

This, then, is the first role that propositions are expected to play: they are the objects of propositional attitudes—such as belief. I turn now to the second role.

2.2.2 Propositions are the contents of declarative utterances

The second role that propositions play is to be the content of declarative utterances. An 'utterance' is typically a token sentence uttered in a particular context: it might be a particular squiggle on a page or a collection of sound waves.[6] An utterance is thus a concrete object. Sometimes the sentences that we utter are 'declarative': that is, they say that something is the case. For example, 'grass is green' is a declarative sentence, whereas 'is grass green?' is not. When a declarative sentence is uttered, it will count as an 'assertion' provided that it is uttered with 'assertoric force'. Giving a precise definition of assertoric force is difficult, but we can get the rough idea through examples. If I say 'grass is green' because I am trying out how the words

---

[6] I say 'typically' because some utterances do not seem to involve sentences at all: a gesture, for example, might be seen as an utterance.

THE ROLES THAT PROPOSITIONS PLAY   11

sound and wondering if this sentence counts as an example of alliteration, then I am not uttering it with assertoric force. Similarly, if I say 'grass is green' because it is the punchline to a joke or a line in a play, then again I'm not uttering it with assertoric force. But if you ask me what colour grass is, and I respond in all seriousness with 'grass is green', then I have made an assertion.

Let us suppose then, that Bertie utters the following sentence with assertoric force—so his utterance counts as an assertion:

'The apple is in the bowl.'

What is the meaning, or content, of Bertie's utterance? On the standard view, we would say that Bertie's utterance expresses a proposition—a certain way that things could be—and this proposition is the content or meaning of his utterance. You might wonder why we don't just say that the content of the utterance is the sentence. If by this we mean the token sentence—the string of noises that Bertie made—then this would be to say that the content of the utterance is simply itself. This doesn't seem right: the utterance represents something—and what it represents is not just itself. Another idea is that the content of the utterance is the sentence type of which Bertie's utterance is a token. But that doesn't seem right either. We have already seen that the same sentence type can be used to make utterances with different contents, as when Bertie says 'The apple is in the bowl' indicating one bowl, and Bobbie says 'the apple is in the bowl' indicating a different bowl. If the content of an utterance is the relevant sentence type, then these two utterances would have the same content. But intuitively Bertie and Bobbie are not saying the same thing: they are not agreeing with each other. By introducing propositions, we can make sense of what is happening here. Bertie and Bobbie are using the same sentence type to express different propositions, and it is these propositions that are the contents of their assertions.

One reason, then, why it's a problem to say that the contents of utterances are sentence types, is that two utterances of the same sentence type can have different contents—as in the case of Bertie and Bobbie's utterance of 'the apple is in the bowl' above. Another problem is that two utterances of a different sentence type can have the same content. For example, if Bertie says 'The apple is in the bowl', and Bertha says 'The bowl has the apple in it', both talking at the same time in the same context, thinking of and pointing to the same apple and bowl, then their utterances have the same content:

12 PROPOSITIONS

Bertha's utterance does not add anything that Bertie has not already said, and vice versa. The case seems even more compelling when we think of sentences that are translations of each other. If (in the same context, with similar intentions) Bertie says 'The apple is in the bowl' and Bertha says 'La pomme est dans le bol', then their utterances have the same content: they mean the same thing, despite the fact that they are token utterances of two different sentence types. Again, if we introduce propositions, then we can explain this: two utterances of different sentence types can express the same proposition, and it is the proposition expressed that is the content of a declarative utterance.

2.2.3 Propositions are truth-apt

To say that propositions are truth-apt is to say that they are the sorts of things that can be true or false. They may not be the only things that can be true or false, for we might also want to say (derivatively, perhaps) that utterances and beliefs can also be true or false. Some would also claim that there are some propositions without truth-values: vague propositions, perhaps, or propositions that are unverifiable. I don't get into this debate here, for the important point is just that at least some propositions are among the things that can be true or false, and we will focus on these truth-apt propositions.

On the standard view, the truth-value of a proposition does not change over time. This might seem strange. To see why, suppose again that Bertie utters the sentence 'The apple is in the bowl', expressing a proposition. If we imagine that the apple is in the bowl at 10 a.m., but somewhere else at 11 a.m., then it might seem that the truth-value of the proposition changes from true to false between 10 a.m. and 11 a.m. This however is not the standard view. On the standard view, if Bertie uttered the sentence 'The apple is in the bowl' at 10 a.m., then he expressed a particular proposition, and the proposition expressed depended on the sentence type and also the context—including (crucially here) the time of utterance. The time of utterance helped to fix which proposition Bertie expressed. We might gloss this by saying that the proposition that Bertie expressed is that there was an apple in the bowl at the time of utterance—i.e. at 10 a.m.[7] This

---

[7] This gloss is not entirely satisfactory—partly for reasons that will become apparent later in the book. I just use this gloss here to introduce the general idea that propositions do not change their truth-values over time.

proposition remains eternally true. At 11 a.m., for example, even if there is by then no longer an apple in the bowl, it is still the case that there was an apple in the bowl at 10 a.m., and so the proposition that Bertie expressed at 10 a.m. remains true. If someone makes another utterance of 'The apple is in the bowl' at 11 a.m., then this utterance will express a different proposition—which again will be either eternally true or eternally false. A proposition then, on the standard view, does not change its truth-value over time.

I have described the three roles that philosophers have standardly expected propositions to play. There are some who think that nothing can play all three of these roles, but there are good reasons to think that something must do (King, Soames, and Speaks, 2014). I assume for much of this book that propositions do indeed play all three of these roles, but my main focus will be on the first role that propositions play: that of being the objects of propositional attitudes. I focus on this role because I think that credences are themselves propositional attitudes, and so it follows that the objects of credence are propositions. Having described the roles of propositions in this section, I move on in the following sections to outline various accounts of propositions.

2.3 Frege's account

I begin with Gottlob Frege, both because his views are of interest in their own right, and because they are the background against which many more recent accounts have been constructed. I start by setting out Frege's analysis of sentences (Frege, 1980 (1879–1903); Beaney, 1996), focusing first on names and predicates.

On Frege's account, names refer to objects: for example, the name 'Anna Mahtani' refers to me. And a predicate can be used to say something about an object. For example, '. . . is tall' is a predicate that can be used to say something about an object (namely that the object is tall).[8] It is a one-place predicate because it is used to say something about a single object. Similarly, '. . . sits', '. . . is a dog', '. . . has a headache' are all examples of one-place

---

[8] I have three dots before 'is tall' because this is the typical way to indicate that a space is left for names (or variables) to be added to a predicate to form a sentence: 'the expression for a function must always show one or more places that are intended to be filled up' (Frege, 1980 (1879–1903), p. 25).

14 PROPOSITIONS

predicates. There are also two-place predicates, such as '...is taller than...', and three-place predicates, and so on. Frege's account can handle all of these different sorts of predicates, but as this is not important for our purposes I will focus just on one-place predicates.[9]

These parts of a sentence (names and predicates) each have what Frege calls a 'Bedeutung', or referent. The referent of a proper name is simply an object—the object that the name refers to. Thus the referent of the name 'Anna Mahtani' is me. The referent of a predicate is slightly more complicated, and it helps to first know about the 'extension' of a predicate. The extension of a (one-place) predicate is the set of all objects that actually fall under that predicate. For example, if we take the predicate '...is tall', then the extension of this predicate is the set of all objects that are tall. With that definition in mind, we can understand Frege's account of predicates: the referent of a one-place predicate (like '...is tall') is a function from objects to truth-values. This function maps every object to a truth-value: to the truth-value *true* if the object is tall (i.e. is in the extension of 'tall'), and to the truth-value *false* if the object is not tall.

If you put a name and a one-place predicate together, then you have a sentence. For example, we can put the name 'Anna Mahtani' and the predicate '...is tall' together to make the sentence 'Anna Mahtani is tall'. Just as the parts of this sentence have referents, so the sentence as a whole has a referent, and on Frege's account the referent of a sentence is a truth-value.[10] Central to Frege's account is the idea that the referent of a sentence is fixed by the referents of its parts. We can see how this works using the sentence 'Anna Mahtani is tall'. The first part of this sentence is a name, 'Anna Mahtani', and the referent of this part is an object (me); the second part of the sentence is a predicate, '...is tall', and the referent of this part is a function that maps objects onto truth-values; when we put these referents together, the object (me) is mapped onto a truth-value, which we can suppose is the truth-value *true*. This gives us the referent of the whole sentence. Thus the referent of the name and the referent of the predicate together determine the referent of the whole sentence.

We can create more complex sentences by combining individual sentences using 'sentential operators'. These sentential operators correspond

---

[9] Frege uses the term 'concept-word' for a one-place function expression, and the term 'relation-word' for an *n*-place function expression, where *n* is 2 or more. Here I just use the term 'predicate' for simplicity of exposition.

[10] It may seem strange to say that the referent of a sentence is a truth-value rather than, say, a state of affairs, and whether Frege was right to say so is debatable (Beaney, 1996).

roughly to terms such as 'and' ($\land$), 'it's not the case that' ($\neg$), and so on. Each of these sentential operators comes with a rule that states how the referent of the new more complex sentence containing the operator is determined by the referents of the simpler sentences that are connected by the operator. For example, if we join two simple sentences using the sentential connector 'and', then the rule is that the new more complex sentence has the referent *true* if and only if[11] both simple sentences have the referent *true*; otherwise the new more complex sentence has the referent *false*. There are many further extensions to Frege's account of reference—such as his account of quantifiers—and Frege's work in this area gives us the fundamentals of contemporary logic. Here I have presented just a basic introduction to his account of reference.

In giving his account of reference, Frege hit a series of problems. Here I focus on the problem that is most relevant to this book. To illustrate the problem, take the following pair of sentences—and let us assume that the sentences are uttered in the same context, with 'Tom' referring to the same particular individual in each:

(2d)　Tom believes that George Orwell is a writer.

(2e)　Tom believes that Eric Blair is a writer.

On Frege's account, the referent of a whole sentence (i.e. its truth-value) is fixed by the referents of the parts of the sentence. And so sentences (2d) and (2e) ought to have the same referent, because each corresponding part of the two sentences has the same referent. For example, the first part of sentence (2d) is the name 'Tom', whose referent is a particular person; sentence (2e) also begins with the name 'Tom' with the same referent. The only words that differ between the two sentences are 'George Orwell' in sentence (2d) and 'Eric Blair' in sentence (2e), but these two different names have the same referent: 'Eric Blair' and 'George Orwell' are two names for the same person. Given that the two sentences are constructed from parts with the same referents, and given that the referent of a sentence is fixed by the referents of its parts, the two sentences ought to have the same referent. That is, on Frege's account, they ought to have the same truth-value. But they may not have the same truth-value: it might be that Tom does not know that Eric Blair and George Orwell are the same person, and so perhaps though Tom

---

[11] For the remainder of this book, I will use 'iff' to mean 'if and only if', as is common practice in philosophical writing.

16 PROPOSITIONS

believes that George Orwell is a writer, he does not believe that Eric Blair is a writer. Thus (2e) can be false while (2d) is true.

The truth-value of sentence (2d) above is sensitive to how George Orwell is designated. If we substitute a different designator—even while keeping the referent the same—we may be able to change the truth-value of the sentence. Philosophers say that here the verb 'believes' has created an 'opaque' or 'intensional' context. Where we have a 'transparent' or 'extensional' context, we can substitute names and predicates freely—without any fear that the truth-value of the sentence will be affected—so long as we keep the referents the same. For example, take the sentence 'Mount Everest is over 8 km tall': the truth-value of this sentence will be unaffected if we substitute 'Chomolungma' for 'Mount Everest' (given that these are two names for the same mountain). We might say that the name in this sentence is referring transparently: its role is to pick out the relevant object, and it doesn't matter (as far as the truth of the sentence is concerned) *how* it does that. In contrast, in sentence (2d) above, the verb 'believes' introduces an opaque or intensional context. The truth-value of the sentence may change if we substitute a different name for 'George Orwell'—even if it refers to the same person.

There are various ways to create an opaque context, and using the psychological verb 'believes' is one such way. Other psychological verbs can also create opaque contexts, including 'desires', 'hopes', and so on. For example, there is a reading on which sentence (2f) below can be true while sentence (2g) is false:

(2f) Tom hopes to meet George Orwell.

(2g) Tom hopes to meet Eric Blair.

In sentences like these which exhibit opacity, it seems that the referent of the sentence is not fixed by the referents of the parts, because we may be able to change the referent (i.e. the truth-value) of the sentence by substituting one of the parts of the sentence for another part with the very same referent.

This was one of Frege's puzzles and, as he recognized, it seems to show that something is amiss with his account of reference. Frege responded to these puzzles by introducing 'senses'. The idea is that just as each part of a sentence has a referent, so it also has a sense, and the sentence as a whole has a sense which is fixed by the senses of its parts. The sense is the 'mode of determination' of the referent. Thus, for example, the terms 'George Orwell' and 'Eric Blair' have the same referent, but they have different senses:

FREGE'S ACCOUNT 17

though they both refer to the same person, they determine this referent in different ways. In general, two terms which have the same referent may have different senses, but the converse is not true: two terms with the same sense must have the same referent. The sense of a term fixes its referent, but the referent of a term does not fix its sense, precisely because there can be many different ways of determining the same referent.

How does Frege's account of sense handle sentences like (2d) and (2e)? Frege can claim that while all corresponding parts of the two sentences have the same referents, they do not all have the same senses: in particular, 'George Orwell' in (2d) and 'Eric Blair' in (2e) have different senses. Thus the sentences as a whole can have different senses. But this does not yet solve Frege's problem, for the *referents* of sentences (2d) and (2e) are also different: sentence (2d) is true, we are supposing, while sentence (2e) is false, and on Frege's account the referent of a sentence is its truth-value. How can two sentences have different referents, if all the parts have the same referents, given that the referent of a sentence is fixed by the referents of its parts? Frege dealt with this problem by claiming that within the context of psychological verbs, words do not have their customary referents: rather, their referents in these contexts are what we would ordinarily consider their senses. Thus, as it appears in sentence (2d), the referent of 'George Orwell' is not the man himself, but rather what would usually be the sense of the term—the mode of determination of the man. And as it appears in sentence (2e), the referent of 'Eric Blair' is this other different mode of determination of the man. Thus as they appear in sentences (2d) and (2e), the terms 'George Orwell' and 'Eric Blair' have different referents, and this allows Frege to explain how the two sentences can have different referents (truth-values), while maintaining that the referent of a sentence is fixed by the referents of its parts.

Let us now consider what propositions are on Frege's account. Here I am focusing on the role of propositions as the objects of propositional attitudes—so the question can be put as follows: on Frege's account, what are the objects of, say, belief? Frege's answer is that the objects of belief are *thoughts*, which are the senses of sentences. When I say that Tom believes that George Orwell is a writer, I am saying that Tom takes to be true the thought which is the sense of the sentence 'George Orwell is a writer'. To say that Tom believes that Eric Blair is a writer is to say that Tom takes to be true a different thought—for the senses of 'George Orwell is a writer' and 'Eric Blair is a writer' are different. This explains why (2d) can be true while (2e) is false.

18 PROPOSITIONS

Introducing senses gave Frege a way to deal with various puzzles that arose for his account of reference—including the puzzle that is most relevant to this book, which is how (2d) and (2e) can have different truth-values. However, it is rather unclear what senses are supposed to be, the key issue being that Frege gives conflicting criteria for sameness of sense (Beaney, 1996). There is no universally accepted account of sense, and I don't attempt to give such an account here. My aim in this section has just been to give a quick overview of Frege's account so that in future chapters we can refer back to it and consider an analogous account of the objects of credence.

2.4 Russell's account

Russell's account can be seen in many ways as a development of Frege's, but one crucial difference is that Russell does not invoke senses. As we saw, Frege claimed that in a sentence such as (2d) the terms do not all have their customary referents:

(2d) Tom believes that George Orwell is a writer.

On Frege's view, the referent of 'George Orwell' in sentence (2d) is not the man himself, which is the usual referent of the name, but rather whatever sense the name would ordinarily have. Russell does not take this view. For Russell, any genuinely proper name always has as its referent simply the object to which it refers. The complication is that for Russell our natural language contains very few genuinely proper names. On Russell's view, 'George Orwell', for example, is only an apparently proper name: once it is correctly analysed it turns out to be something very different.

To understand how Russell analyses the apparently proper names in our natural language, we first need to understand his account of definite descriptions. Take this sentence:

(2h) The author of 1984 is tall.

This sentence contains a definite description: 'the author of 1984'.[12] On Frege's view, sentence (2h) can be broken into parts, each with its own

---

[12] Definite descriptions typically start with 'the' (e.g. 'the cat I saw this morning'), whereas indefinite descriptions start with 'a' (e.g. 'a cat I saw this morning').

sense and reference in the usual way. The description 'The author of 1984' functions just like a proper name, having both a referent and a sense; similarly, '... is tall' is a predicate with a referent and a sense. For Russell, however, the definite description 'the author of *1984*' is not like a proper name at all: the real logical form of the sentence turns out to be quite different on analysis. Sentence (2h) on Russell's view can be analysed as the conjunction of (2h.i)–(2h.iii) below (Russell, 1905):

(2h.i)   There exists an author of *1984*.

(2h.ii)   There exists at most one author of *1984*.

(2h.iii) Everything that is an author of *1984* is tall.

We have seen, then, how Russell analyses sentences containing definite descriptions. How does this relate to apparently proper names such as 'George Orwell'? The answer is that on Russell's account 'George Orwell' is really a definite description in disguise (Russell, 1905). Exactly which definite description corresponds to 'George Orwell' is not at all obvious. Is it the description that we all associate with the name 'George Orwell'? But plausibly we do not all associate the same description with that name, and perhaps some of us—even competent users of the name—do not have *any* particular description associated with the name (Kripke, 1980). Let's put these issues to one side for now, and just suppose that 'George Orwell' corresponds to this description: 'the person who took up the pen name "George Orwell" and wrote *1984*'. And let's suppose that 'Eric Blair' corresponds to a different description: 'the person who was christened "Eric Blair", and who was born on 25 June 1903'. Then we can reconsider the sentences below:

(2d)   Tom believes that George Orwell is a writer.

(2e)   Tom believes that Eric Blair is a writer.

Following Russell, we can give a partial analysis of (2d) and (2e) as follows:

(2d')   Tom believes that the person who took up the pen name 'George Orwell' and wrote *1984* is a writer.

(2e')   Tom believes that the person who was christened 'Eric Blair', and who was born on 25 June 1903 is a writer.

20   PROPOSITIONS

There are lots of differences between the sentences (2d') and (2e'), and no reason to think that the referents of the terms in (2d') all match up to the referents of the terms in (2e'). Thus we can maintain the view that the referent of a sentence—that is, its truth-value—is fixed by the (standard) referents of its parts, while holding that (2d') and (2e')—and so (2d) and (2e)—have different truth-values. Thus Russell can explain how it is that (2d) and (2e) have different truth-values, without any need to introduce senses.

Now let's consider what propositions are on Russell's account. That is, what are the objects of propositional attitudes like belief? We have seen that for Frege the objects of belief are *thoughts*, and these are the senses of sentences—fixed by the senses of the parts of the sentence. For Russell the constituents of a proposition are objects and properties.[13] These objects and properties are the referents of the parts of the sentence—once the sentence has been analysed to reveal its true logical form. The referent of a true proper name—that is, a *logically* proper name—is an object. On Russell's account there are not many logically proper names in natural language. As we have seen, Russell maintains that most ordinary proper names are definite descriptions in disguise—but in English the terms 'this' and 'that' count as logically proper names, and so (when uttered in some specific context) these terms have objects as referents. The other parts of a sentence, when properly analysed, refer to properties. On Russell's view, these objects and properties are the constituents of a proposition, but a proposition is something over and above the collection of its constituents: the proposition also has some sort of structure.

The details of Russell's account are not entirely clear, but many more recent philosophers have developed accounts of structured propositions that build on Russell's account while combining it with the idea of 'guises', and I turn to these accounts now.

2.5 Guise Russellianism

Several philosophers have taken up Russell's idea that propositions are structured entities: advocates of this sort of view include (Soames, 1987;

---

[13] More precisely, the constituents are objects and relations: one-place relations (or properties) corresponding to one-place predicates, two-place relations corresponding to two-place predicates, and so on. For simplicity, I just talk about objects and properties in the main text.

Salmon, 1986). These philosophers do not follow Russell in claiming that there are very few logically proper names in natural language, and so they need a different response to the problem discussed above—that (2d) and (2e) can have different truth-values. To respond to that sort of problem, these philosophers supplement their position by introducing the idea of 'guises', and so we can call the resulting view 'guise Russellianism'.

On this view, a sentence such as 'Fred is taller than Frank' contains names and predicates, and these have 'semantic values', which are the referents of the names and predicates. Guise Russellians would treat 'Fred' and 'Frank' as logically proper names (rather than as definite descriptions in disguise), and so the semantic values of these two names are simply the two people referred to. The semantic value of '...is taller than...' is a relation (similar to a property). The proposition expressed by the whole sentence consists of these semantic values structured in a certain way. The sentences 'Fred is taller than Frank' and 'Frank is taller than Fred' express different propositions, because though the propositions contain all the same parts (that is, the same people and the same relation), they are differently structured.

On the structured proposition account, 'George Orwell is a writer' and 'Eric Blair is a writer' express the same proposition: this proposition consists of an object (which can be denoted either by 'George Orwell' or 'Eric Blair'), and a property denoted by the other parts of the sentences, arranged in a particular structure. The problem is that, as we've already seen, it seems obvious that Tom might believe that George Orwell is a writer without believing that Eric Blair is a writer. But how can he both believe and not believe the very same proposition? Nathan Salmon responds to this dilemma by denying the claim that I said 'seems obvious'. On Salmon's view, if Tom believes that George Orwell is a writer, then he also believes that Eric Blair is a writer (and vice versa). This might seem surprising, but there is a sense in which it is intuitively right: if Tom believes that George Orwell is a writer, then in a way he *does* believes that Eric Blair is a writer.

Salmon maintains that belief is a two-place relation between a person and a proposition: for example, Tom stands in the belief relation to the proposition (call it P) that can be expressed either by 'George Orwell is a writer' or 'Eric Blair is a writer'. But Salmon also introduces a three-place relation, 'BEL', which can hold between a person, a proposition, and a *guise*. A guise is a mode of determination of a referent—and in some ways it is rather like Frege's 'sense'. The BEL relation holds between Tom, the proposition P, and the 'George Orwell' guise, but the BEL relation doesn't hold between Tom, P, and the 'Eric Blair' guise. On Salmon's view, the belief relation holds

22 PROPOSITIONS

between a person and a proposition provided that the BEL relation holds between that person and proposition and *some* guise. Thus it is true to say that Tom believes that George Orwell is a writer, and equally true to say that Tom believes that Eric Blair is a writer. To account for our intuition that there is something wrong with saying that Tom believes that Eric Blair is a writer, Salmon claims that to say that Tom believes that Eric Blair is a writer entails that Tom stands in the BEL relation to the proposition expressed under some guise or other, but also implies (without entailing) that the relevant guise is the 'Eric Blair' guise. This implication is false, and so the claim may be misleading, while being strictly true.

I explore Salmon's account in more depth in chapter 4, but for now I turn to an alternative account of propositions: possible world semantics.

2.6 Possible world semantics

Along with propositions, possible worlds are another idea that philosophy students encounter at an early stage. A possible world is a 'way things could have been'. Usually we take possible worlds to be complete, in that each possible world is a way things could have been down to every last detail. If you could have been a lead vocalist in a band, then there is a possible world where you are a lead vocalist in a band—and indeed there are multiple such possible worlds exemplifying all the different ways that this could have been true: you could have been a lead vocalist in a band playing at The Apollo this weekend, while owning a tabby cat called 'Martha'—and so on. The number of possible worlds is obviously vast, and arguably infinite. There is debate over whether possible worlds are real in any sense, or whether they are just convenient fictions (Lewis, 1973). I do not get into this question in this chapter, as all that I say should make sense whichever view you take. If you think of possible worlds as convenient fictions, then you can understand any talk of possible worlds as shorthand for talking about what could have been the case.

Propositions can have different truth-values at different possible worlds. For example, the proposition that you have a tabby cat may be true at one possible world but false at another, and so the truth-value of this proposition is relative to a possible world. When we say simply that a proposition is true or false, we typically mean that it is true or false at the actual world. Some propositions are true at all possible worlds, and these are the *necessary* propositions. A proposition is *possible* iff it is true at some possible world.

A proposition that is possible but not necessary—i.e. that is true at some but not all possible worlds—is *contingent*.

Philosophers usually think that possibility comes in lots of different flavours. We might say that it is not possible for humans (constructed as they are) to fly unaided—meaning that this is not *physically* possible: roughly, it is not possible given the laws of physics. Similarly, we can think of propositions that are not biologically possible, psychologically possible, legally possible, practically possible, and so on. We might use the possible worlds framework to model these sorts of possibility by focusing on just those possible worlds where the laws of the relevant discipline are not violated. There is also a category of *impossible* worlds, and I return to this in chapter 8.

Philosophers also talk about *epistemic* possibility. Sometimes by this they mean possibility compatible with what is known—and as, of course, what is known varies from person to person and time to time, this sort of possibility will be relative to an agent at a time. For example, for me right now it is not epistemically possible that London is in France, because I *know* that it isn't, but that proposition may be epistemically possible for someone else. Sometimes philosophers use the idea of epistemic possibility a bit differently, connecting it to the distinction between a priori and a posteriori knowledge. A proposition is knowable a priori if it can be known without requiring any particular evidence. Examples of propositions that can be known a priori are logical propositions (e.g. the proposition expressed by 'if London is in France then London is in France'), and mathematical propositions (e.g. $2 + 2 = 4$). Other propositions (e.g. the proposition that London is not in France) are only knowable a posteriori—that is, after gaining some evidence, whether from observation, testimony, or some other source. Some philosophers model this distinction between a priori and a posteriori propositions using a possible worlds framework, with a set of epistemically possible worlds at which all a priori knowable propositions obtain.

There is another sort of possibility which is particularly important to philosophers: metaphysical possibility. The set of metaphysically possible worlds is wider than the set of physically possible worlds: some propositions are physically impossible but metaphysically possible. For example, it is physically impossible for humans to fly unaided, but it is metaphysically possible: though there is no physically possible world where humans fly, there is a metaphysically possible world where they do. Exactly where the boundary lies to metaphysical possibility is a controversial matter. On some

24 PROPOSITIONS

views, if some proposition is conceivable then it is metaphysically possible—but this claim is debated, and there is further disagreement over what it is for a proposition to be conceivable (Yablo, 1993). It is also unclear whether metaphysical possibility is the most permissive form of possibility, or whether there are yet more permissive forms of possibility. Perhaps there are propositions which are logically possible without being metaphysically possible? For example, take the proposition expressed by 'this ball is red all over and green all over': no logical law is violated in this sentence, and yet the proposition expressed seems inconceivable, and—in some sense—impossible. It may be then that there are possible worlds—logically possible worlds—beyond those that are metaphysically possible. Furthermore, many theorists hold that there are propositions which are not knowable a priori (and so that are false at some epistemically possible world) while being metaphysically necessary—and this provides another reason to think that there are possible worlds beyond those that are metaphysically possible (Kripke, 1980; Edgington, 2004). We will revisit these questions in chapter 8, but for the rest of this chapter I will use 'possible world' as it is most typically used by philosophers—to mean metaphysically possible world.

Some philosophers (e.g. Carnap, 1947; Kripke, 1959) use possible worlds to give an account of propositions—calling this 'possible world semantics'—and I turn to this now. I begin by recalling that on Frege's account a predicate has an 'extension', which is the set of objects that fall under that predicate. Two apparently different predicates can have the same extension. For example, it might just so happen that all existing creatures that are renate (i.e. have a kidney), are also cordate (i.e. have a heart). Then the predicates '... is cordate' and '... is renate' have the same extension. And yet (we might think) these two predicates don't have the same content, and so it follows that the content of a predicate can't be captured just by its extension. In response, we can point out that though as it happens all renate creatures are cordate, this isn't *necessarily* the case: it could have been otherwise. That is, there are possible worlds where there are renate creatures that are not cordate, and vice versa. Perhaps, then, the content of a predicate can be captured by thinking about the objects that fall under it *at each possible world*. We can say that the content of a predicate can be given by its 'intension', which is a function from possible worlds to sets of objects. Though the extension of '... is cordate' is the same as the extension of '... is renate', the *intensions* of these predicates are different—and this reflects the fact that they have different contents.

Similarly, we can think of designators—names and definite descriptions—as having both extensions and intensions. Take the definite descriptions 'the person who was christened "Eric Blair" and who was born on 25 June 1903', and 'the winner of the 1946 Hugo Prize'. As it happens, these two definite descriptions have the same extension—for the person (George Orwell) who was christened 'Eric Blair' and who was born on 25 June 1903 also won the 1946 Hugo Prize. But this does not hold across all possible worlds: there are worlds where someone else won the prize instead. The intensions of the two descriptions are functions from possible worlds to people: they happen to both map the actual world onto the same person (George Orwell), and so they have the same extension, but they don't map identically across all possible worlds, and so they have different intensions. In general, the intension of a definite description maps a possible world onto whoever happens to fit that description at that world.

In contrast, consider a proper name, such as 'George Orwell'. The extension of this name is a particular person. What is the intension of this name? Typically proper names are taken to be 'rigid designators', picking out the same object at every possible world (Kripke, 1980), and so the intension of the name is a function from each possible world (where George Orwell exists) to the very same person (George Orwell). To claim that a name designates rigidly is not to claim that people at all possible worlds use the string of letters that make up the name in the same way. We can imagine a possible world where the string of letters 'George Orwell' is used to mean 'Hello!', and when someone at that world says 'George Orwell' the meaning of their utterance is 'Hello!' To claim that 'George Orwell' designates rigidly is not to claim that this string of letters has the same meaning regardless of where it is uttered. Rather, the claim is that as uttered by us—with its actual meaning—'George Orwell' picks out the same person at every possible world (where that person exists).

Here is an example to help explain and motivate this distinction between definite descriptions and proper names. First consider whether George Orwell could have been a woman. In other words, is there a possible world where George Orwell is a woman? This is debatable: it seems to depend on whether George Orwell's gender is part of his 'essence' (as some would claim), or whether he could have had a different gender and yet still have been the same person (as many others would maintain). But the important point to notice here is that the name 'George Orwell' is designating rigidly: the question is over whether there is a possible world where

26 PROPOSITIONS

George Orwell—*that very person*—is a woman.[14] In contrast, consider whether the winner of the 1946 Hugo Prize could have been a woman. In other words, is there a possible world where the winner of the 1946 Hugo Prize is a woman? This is relatively uncontroversial: the judges could have picked a different book—Agatha Christie's *The Hollow*, for example—and then a woman would have won the 1946 Hugo Prize. A possible world where the winner of the 1946 Hugo Prize is a woman isn't necessarily a possible world where George Orwell (the *actual* winner of the 1946 Hugo Prize) is a woman—for 'the winner of the 1946 Hugo Prize' refers to different people at different possible worlds. In this way we can see that names and definite descriptions behave differently: names designate rigidly, whereas definite descriptions have a non-rigid reading.[15] And so there is a corresponding difference in their intensions: the intension of a name maps each possible world (where the relevant person or object exists) to the same person or object, whereas the intension of a definite description can map different possible worlds to different people or objects. Thus a name and a definite description (or two definite descriptions) might pick out the same object in the actual world, and so have the same extension, but have different intensions. But any two names with the same extension (i.e. that pick out the same object in the actual world) will also have the same intension: for example, the intension of 'George Orwell' is the same as the intension of 'Eric Blair'.

Here I move on to define extensions and intensions for whole sentences. The extension of a sentence is its referent, which (as Frege claimed) is its truth-value, and its intension is a function from possible worlds to truth-values. Thus, for example, take the sentence 'grass is green'. This sentence is true at the actual world, and so its extension is the truth-value *true*—the

---

[14] If we can talk about how things are with George Orwell at other possible worlds, does it follow that there must be some unique and essential features of George Orwell by which we could recognize him? That doesn't follow. As Kripke writes: ' "Possible worlds" are stipulated, not discovered by powerful telescopes' (Kripke, 1980, p. 262). When we describe a possible world, we can use names in the description, and when we do so we are *stipulating* how things are at that world with the individual named—the individual that bears that name in the actual world. Of course it can still be open to debate (as in the example in the main text) whether the world stipulated is possible.

[15] Names are typically taken to have *only* a rigid reading, whereas in many contexts there are two possible readings of a definite description: one on which it designates rigidly, and one on which it designates non-rigidly. A definite description can, however, be *rigidified* by adding an 'actually' operator, and then—like a proper name—it has only a rigid reading. For example, the definite description 'the *actual* winner of the 1946 Hugo Prize' is a rigid designator. I discuss rigidified definite descriptions again in section 7.2.

same as the extension of 'snow is white'. In considering the intensions of these sentences, we think about their truth-values at various other possible worlds. The intension of 'grass is green', for example, is a function that maps those worlds at which grass is green onto *true*, and all other worlds onto *false*. There are worlds where grass is green but snow is not white, and vice versa, so though the sentences 'grass is green' and 'snow is white' have the same extension (i.e. they have the same truth-value at the actual world), they have different intensions. According to many supporters of possible worlds semantics, the content of a sentence—i.e. the proposition that it expresses—is its intension. Thus a proposition can be understood as a function from possible worlds to truth-values, or more simply just as a set of possible worlds—the possible worlds mapped to *true*. This approach fits very naturally into the credence framework, as we shall see in future chapters, though it also faces problems.

To see one problem which is particularly important for this book, we can start by comparing the two sentences 'The winner of the 1946 Hugo Prize is a writer' and 'George Orwell is a writer'. These sentences have the same extension, for both are true, but they have different intensions. For example, there is a world where George Orwell is not a writer, and someone else who is a writer wins the 1946 Hugo Prize: the intension of 'the winner of the 1946 Hugo Prize is a writer' will map this world to *true*, but the intension of 'George Orwell is a writer' will map this world to *false*, and thus these two sentences have different intensions. In contrast, the two sentences 'George Orwell is a writer' and 'Eric Blair is a writer' have the same intensions: any world where George Orwell is a writer is also a world where Eric Blair is a writer, and vice versa. This is because 'George Orwell' and 'Eric Blair' are rigid designators, picking out the very same person at every possible world. This creates a prima facie problem for possible world semantics, for if the proposition expressed by 'George Orwell is a writer' is the same as the proposition expressed by 'Eric Blair is a writer', then how come Tom can believe one but not the other, as (2d) and (2e) imply? This is a problem that we will return to throughout this book.

Before leaving this brief overview of possible worlds semantics, I clarify a key distinction between this account and Russellianism/guise Russellianism, and also mention one influential development of the framework. I begin with the key distinction between this account and Russellianism/ guise Russellianism, which is that according to Russellianism/guise Russellianism propositions are structured entities, whereas possible

28   PROPOSITIONS

world semantics does not take propositions to be structured in the same way.[16] To see the contrast here, take the following four sentences:

(2i)   Tom is taller than Tim.

(2j)   It's not the case that it's not the case that Tom is taller than Tim.

(2k)   Tim is shorter than Tom.

(2l)   Tom is taller than Tim, and $2 + 2 = 4$.

Each of these sentences is differently structured, and according to Russellianism/guise Russellianism each expresses a different proposition. For example, the proposition expressed by (2l) contains the number 2, and this number is not a part of the proposition expressed by (2i), and so the propositions expressed by (2i) and (2l) are distinct. In contrast, given possible worlds semantics each of (2i)–(2l) expresses the same proposition, for each has the same intension: the set of possible worlds where (2i) holds is exactly the same as the set of possible worlds where (2j) holds, and so on for (2k) and (2l).[17] According to possible worlds semantics, a proposition corresponds to a set of possible worlds, and this is not structured—at least not in the way that Russellians and guise Russellians take propositions to be structured. This is a key distinction between these different sorts of accounts, and I will refer to it in places throughout the book.

Finally, I turn to an influential development of the possible worlds semantics framework. The innovation is to replace possible worlds with centred possible worlds (Lewis, 1979). A centred possible world is a standard metaphysically possible world plus a centre, where that centre is typically but not universally taken to be an individual and a time (Liao, 2012). To see why it seemed like a good idea to introduce these worlds, we can consider Perry's example of a person with amnesia wandering around Stanford library (Perry, 1977). This person reads a biography about Rudolph Lingens, and at some point comes to realize that *he* is Rudolph Lingens. What proposition has he just learnt? According to possible worlds semantics, a proposition corresponds to a set of possible worlds, but what set of possible worlds corresponds to the proposition that Lingens has just learnt? The problem here is that possible worlds are generally thought of

---

[16] Arguably Frege also took propositions to be structured, though this is open to debate.

[17] This holds on a typical possible worlds framework, but other variations are possible. I explore some of these variations in chapter 8.

as *objective*, and what Lingens has learnt is something that is not objective, but true from his own perspective. By introducing centred possible worlds, we seem able to handle this sort of case: the proposition that Lingens has learnt is true at every centred possible world where the person at the centre is Lingens. Thus we should see propositions as functions from centred possible worlds to truth-values—or more simply just as sets of centred possible worlds.

This new account of propositions takes some getting used to. If I say 'I am hungry', and you say 'I am hungry', are we expressing the same or different propositions? Assuming possible worlds semantics with centred possible worlds, we are expressing the very same proposition—which corresponds to the set of centred possible worlds with a hungry person at the centre. In some ways this seems like the right thing to say, because if we both believe what we say then we will behave in a broadly similar way. But in another sense it seems like the wrong thing to say, for your assertion might be true while mine is false—and how can this be if we are both expressing the same proposition? To get my head around this idea, I have found it helpful to remember that on this account you and I are at different worlds—different centred possible worlds, that is. I am at a world with myself at the centre, and you are at a world with yourself at the centre. It is generally understood that the truth-value of a proposition can vary depending on which world it is evaluated at. For example, the proposition that grass is green is true at the actual world, but there are possible worlds where this proposition is false. So we can already understand the idea that two people can express the same proposition, one truly and one falsely, provided that the people are at different possible worlds. And it is just the same phenomenon on the centred worlds account: you and I can both express the same proposition with our utterance of 'I am hungry', and our utterances can differ in truth-value because we are at different (centred) possible worlds. Similarly, as time passes we are each continually moving between worlds—from one centred world to another—and so the proposition expressed by 'it is now 4 p.m.' can be false at one time and true at another. We will explore the idea of centred possible worlds further in chapter 7.

2.7 Chapter summary

Philosophers generally hold that propositional attitudes, such as beliefs and desires, are attitudes towards propositions. Propositions are usually also

30 PROPOSITIONS

assumed to play other roles too: they are the content of declarative utterances, and they are truth-apt.

There is widespread disagreement over what propositions are. Here I have described some of the accounts in the literature, including Frege's account on which propositions are understood in terms of senses; Russell's account on which propositions are structured entities; guise-based versions of accounts of structured propositions; and a version of possible worlds semantics. Each of these accounts face various problems, and we will be revisiting many of them throughout the book.

This book brings work in the philosophy of language together with the credence framework. Having introduced the relevant material from the philosophy of language in this chapter, I turn in the next chapter to the credence framework.

3
The Credence Framework

3.1 Introduction

This book brings together two different but related areas of research: accounts of propositions and propositional attitudes found in the philosophy of language; and the credence framework, as used in a wide variety of disciplines. In this introductory part of the book, I am giving an introduction to each area, aiming to equip readers to understand the connections that I will draw later. The last chapter (chapter 2) introduced the relevant material from the philosophy of language, and in this chapter (chapter 3) I introduce the credence framework.

The basic idea behind the credence framework is quite intuitive. In the last chapter I touched upon belief—standardly understood as a relation between a person and a proposition. There we assumed that belief is an all or nothing matter: either you believe a proposition or you don't. But intuitively there are finer distinctions to be drawn. For example, you might be pretty sure that it will rain at some point tomorrow, but even more sure that it will rain at some point in the coming year. To say that you simply believe both of these propositions is to miss the difference between your epistemic attitudes towards each. We might think, then, that a person's epistemic attitude towards a proposition is not really captured by simply stating whether or not she believes that proposition, but might be better captured by giving her *degree of belief* or *credence* in a given proposition. On the standard view, a person's credence in a proposition is given by a number between 0 and 1, with 1 representing certainty in the proposition, 0 representing absolute disbelief, and the numbers between 0 and 1 representing the gradation of attitudes between certainty and absolute disbelief.

This is the intuitive idea behind the framework, but I turn now to some specifics.

The Objects of Credence. Anna Mahtani, Oxford University Press. © Anna Mahtani 2024.
DOI: 10.1093/oso/9780198847892.003.0003

32 THE CREDENCE FRAMEWORK

3.2 The probability framework

I begin by describing the probability framework.[1] On this framework, we can construct various 'probability spaces', which are abstract, mathematical objects. These objects can then be interpreted in various ways—and on one such interpretation we use probability spaces to represent peoples' epistemic states. But we begin with the probability framework in the abstract.

We start with a finite set of 'states', and we will call this set Ω. We can label the states that Ω contains $w_1, w_2, \ldots w_n$. So $\Omega = \{w_1, w_2, \ldots w_n\}$. We can consider various subsets of set Ω, such as $\{w_1\}$, $\{w_1, w_2\}$, and so on. If we join two or more of these subsets together, then we get unions: for example, the union of $\{w_1\}$ and $\{w_2\}$—written $\{w_1\} \cup \{w_2\}$—is $\{w_1, w_2\}$. The union of all the subsets in Ω is Ω. We can also consider the complements: the complement of a subset of Ω is a subset that contains all the *other* states in Ω. For example, the complement of subset $\{w_1\}$—written $\{w_1\}^C$—is the subset $\{w_2, w_3, \ldots w_n\}$. The complement of Ω is the empty set $\varnothing$. Finally, we can consider the intersections: the intersection of two sets is the set containing the states that they have in common. For example, the intersection of $\{w_1, w_2\}$ and $\{w_1\}$—written $\{w_1, w_2\} \cap \{w_1\}$—is $\{w_1\}$.

We can then define an 'algebra' over Ω. This is a set of subsets of Ω, and it is required to have certain features to qualify as an algebra: it must contain Ω, and it must be 'closed under complement and union'. This means that for any set in the algebra, its complement must be in the algebra too; and for any collection of sets in the algebra, the union of those sets must be in the algebra too. So, for example, if $\{w_1\}$ appears in the algebra, then its complement $\{w_2, w_3, \ldots w_n\}$ must also be in the algebra; and if $\{w_1\}$ and $\{w_2\}$ are both in the algebra, then their union $\{w_1, w_2\}$ must be too. From these rules, it follows automatically that $\varnothing$ must also be in the algebra: it must be included as the complement of Ω. Let us call the elements of an algebra 'events' (e).[2]

We can then define a probability function P over our algebra. This function will assign a (single) number to every element of the algebra. For

---

[1] In this chapter I describe one general approach for setting up the probability framework. Another approach has the states in Ω as sentences, a 'Boolean algebra' over these sentences, and the probability function having the elements in the Boolean algebra as its domain. I do not discuss this general approach here because there are problems with the idea of taking the objects of credence to be sentences—problems parallel to the problems discussed in chapter 2 with the idea of taking the objects of belief to be sentences.

[2] For some purposes, we may need an infinite set of states—and then for our algebra we'll need what is called a 'sigma algebra', which is closed under *countable* unions and intersections. For simplicity I focus here on a finite set of states.

THE PROBABILITY FRAMEWORK 33

example, for a given event e in the algebra, the function will assign a number v, and we express this by writing 'P(e) = v'. For P to qualify as a *probability* function it must meet certain conditions. There are various versions of these conditions, but here I will focus on the 'Kolmogorov axioms' (Kolmogorov, 1933 (1950)):

- The probability of an event is a non-negative real number.
- The probability of Ω is 1.
- For any two events e_1 and e_2, if e_1 and e_2 are disjoint (i.e. they have no members in common), then P(e_1) + P(e_2) = P($e_1 \cup e_2$).

From these axioms, all sorts of other facts about the probability function follow. For example, consider any event e and its complement e^C. These events e and e^C are disjoint, and so the numbers assigned to them must sum to the number assigned to their union $e \cup e^C$ (by the third axiom). But the union of an event and its complement is Ω, and the number assigned to Ω is 1 (by the first axiom). Thus P($e \cup e^C$) = 1, and so P(e) + P(e^C) = 1: in other words, for any event e, P(e^C) = 1–P(e). This is one interesting fact about the probability function. And we can use it to derive a further fact: that for any event e in the algebra, the number the probability function assigns to that event is between 0 and 1 (inclusive). To see why, consider first that from the first axiom it follows that the number assigned cannot be below 0. And we have already seen that the number assigned to an event and its complement must sum to 1, and given that no event can be assigned a negative number (from the first axiom), it follows that the numbers assigned to the event and the complement must both be no greater than 1. Thus for any event e in the algebra, $0 \leq$ P(e) ≤ 1. And many further interesting results can be derived (Skyrms, 2000).

A 'probability space' consists of a set of states (Ω), an algebra ($\mathcal{F}$) over that set of states, and a probability function on $\mathcal{F}$ (P). A probability space is thus an abstract mathematical object, and the interest is in how it can be interpreted. There are many possibilities here, but here is a rough overview of a typical interpretation. We see the states in Ω as possible 'outcomes', or ways that the world might turn out to be. We might treat these states as possible worlds, complete in every detail—or at least complete in every detail relative to our interpretative purposes.[3] Then the events in $\mathcal{F}$—which are

---

[3] In practice, states are often taken to be the possible outcomes of a particular experiment rather than entire possible worlds.

34 THE CREDENCE FRAMEWORK

sets of these possible worlds—include more coarse-grained ways that the world might be: we might see these as propositions. And finally, the probability function gives the probability of each of these events, though here there is plenty of room for variation as there are different sorts of probability. I will briefly describe some of these different sorts of probability below, but first here is an example to illustrate (in broad outline) a typical interpretation.

I am about to throw a die. For our purposes, there are six possible outcomes, or states: state 1 (where the die lands with the '1' face up), state 2 (where the die lands with the '2' face up), and so on. Thus $\Omega = \{1,2,3,4,5,6\}$. Then we can define various algebras over Ω—and here is one such algebra: $\mathcal{F} = \{\{1,3,5\}, \{2,4,6\}, \Omega, \varnothing\}$. Effectively, the events in $\mathcal{F}$ are the event of the die landing on an odd side, the event of the die landing on an even side, the event of the die landing on some side or other, and the null event. Now we can introduce a probability function on $\mathcal{F}$. Obviously there are many functions on $\mathcal{F}$ that obey the probability axioms, but here I describe just one such function P: $P(\{1,3,5\}) = 1/2$, $P(\{2,4,6\}) = 1/2$, $P(\Omega) = 1$, and $P(\varnothing) = 0$. This triple—Ω, $\mathcal{F}$, and P—is a probability space, and we are taking the probability function P to give the probabilities of the events in $\mathcal{F}$.

But what exactly are probabilities? My main focus in this book is on subjective probabilities—used to represent a person's epistemic state at a particular time. But before turning to a detailed account of this, I just briefly mention a few other ways that we might understand probability. On one such interpretation, we might think of probability as closely connected with frequency (Venn, 1876; von Mises, 1928 (1957); Reichenbach, 1949). We could take a series of rolls of our die, and consider the proportion of those times that the die landed odd-side up. This gives us, in one sense, the probability of the die landing odd-side up. This leaves many questions to be answered—for example, there is a question over the relevant 'reference class' (should we include just rolls of this die, or rolls of other dice too?), and whether the 'series' of rolls should be some concrete actual series or some theoretical infinite series—and there are different sorts of frequentists who answer these questions in different ways. Alternatively, we might think of probability as objective chance, where the chance of a particular die landing odd-side up can vary over time depending on what is already fixed or determinate at each point (Popper, 1959; Lewis, 1987; Gillies, 2000). I discuss one such version of this interpretation in chapter 5, but there are a range of variations on this interpretation of probability as objective chance—besides alternative accounts of classical (Laplace, 1814 (1951))

WHAT ARE CREDENCES? 35

and logical (Keynes, 1921; Carnap, 1950) probability. Here I set these accounts aside and turn to interpretations on which P represents subjective probabilities.

Subjective probabilities are probabilities that are relative to a person and a time. Suppose that I have just rolled a die and I can see how it has landed, but it is hidden from your view. What is the probability that it has landed odd-side up? Here we might say that the probability *for you* is 1/2, but *for me* it is either 1 or 0, because I can see how it has landed. And if I then reveal the die to you, then the probability *for you* is now either 0 or 1.[4] On this view, two people can have different probabilities in the same event, and a person's probabilities in an event can change over time. Thus we can see a probability space as a way of representing the epistemic state of an individual (an 'agent') at a time. That is, we can adopt the 'credence framework'. On this interpretation, the events are those propositions that the individual can entertain, and the probability function gives her 'credences' or 'degrees of belief' in each proposition. But what exactly are these credences? This is the question that I turn to now.

3.3 What are credences?

One way into this question is to consider how you would go about finding out someone's credence in a proposition. Many theorists see a relation between a person's credences and the way that the person bets—or would bet if given the opportunity. For example, suppose that before we see how the die lands, I offer you a bet whereby I give you £1 iff the die lands odd-side up, and nothing otherwise. How much would you be willing to pay for

---

[4] You might question whether—even after you get to see the die—your credence that it landed odd-side up should be 1. After all, your eyes might be deceiving you, so you shouldn't be *absolutely certain* that the die landed odd-side up based on perceptual evidence. The idea of 'contextualism' may help untangle this issue. Many philosophers hold that the term 'knows' is context dependent: it might be true to utter 'S knows that P' in one context, but false to utter the same sentence (referring to the same individual S, at the same time, and so on) in a different context—simply because the standards for knowledge are higher in the second context than they are in the first. And one way to raise the standards for knowledge is to make salient certain possibilities—such as the possibility that your eyes might not be working properly (Lewis, 1996). A similar phenomenon may hold for credences, so that in one context it may be true to utter 'S has a credence of 1 in P', even though a similar utterance (referring to the same individual S and so on) uttered in a different context is false (Clarke, 2013; Beddor, 2020; Kauss, Forthcoming). For these and other reasons the claim I made in the introduction—that an agent has a credence of 1 in P iff she is *certain* that P—skates over some complexities.

36   THE CREDENCE FRAMEWORK

this bet? If your credence that it will land odd-side up is 0.5, then presumably you'd be willing to pay up to £0.50 for this bet. More generally, take any event e: offer a person a bet whereby she gets £1 iff e and nothing otherwise, and consider the maximum amount that the person would be willing to pay for this bet. We can take this maximum amount to be the person's credence in e.[5]

For some theorists, your (actual or dispositional) betting behaviour *constitutes* your credences: for these theorists, having a particular credence just is betting in a particular way, or being disposed to so bet. Thus we can say that *by definition* a person's credence in e corresponds to the maximum amount she would pay for the bet described above. For other theorists, betting behaviour serves rather as a manifestation or effect of your credences: that your credence in e is some particular value *causes* your willingness to pay up to that amount for the bet described above. In either case—whether causal or constitutive—there is often taken to be a close relationship between (dispositional) betting behaviour and an agent's credences.

But some theorists think that we should broaden our focus from (dispositional) betting behaviour to (dispositional) choice behaviour more generally. One reason to do so is that some people may (for religious or ethical reasons, for example) refuse to bet under any circumstances. In such cases there may be *no* maximum amount that the person would be willing to pay for a given bet—or we might say that the amount is always £0—and yet we may still think that this person has various credences in various propositions. Thus it seems that there is not always a close relationship between credences and betting behaviour after all. We can solve this problem by broadening our conception of what counts as a bet to encompass more general choice behaviour.[6] As Frank Ramsey writes, 'all our lives we are in a sense betting. Whenever we go to the station we are betting that a train will really run, and if we had not a sufficient degree of belief in this we should decline the bet and stay at home' (Ramsey, 1931).

But how can we read off a person's credences from her choice behaviour? Suppose, for example, that the person in Ramsey's situation decides to go to

---

[5] There are other (more careful) ways of designing bets to elicit credences (de Finetti, 1931; Gillies, 2000). The method described in the main text is just a simple example.

[6] Some go still further and deny that we can read off credences from choice behaviour more generally, because a person might have no preferences and make no choices, and yet nonetheless have credences (Eriksson and Hájek, 2007).

the station. Does it follow that her credence that the train will run is, say, above 0.5? Not necessarily! That will depend on how much she values catching the train relative to other possible outcomes. If she is desperate to catch the train, then she might go to the station even if her credence that the train will run is very low; if she doesn't want to catch the train, then she may not go to the station even if she is sure that the train will run. To be able to assess her credence here, we need to know how much she values the various outcomes (catching the train, not catching the train). When we focused on (explicit) bets, it looked as though this measure of the value of outcomes was already supplied by the financial outcomes. I simply assumed that £1 was twice as valuable to you as £1, and so that you'd be willing to risk your £0.50 to get £1 iff your credence that you'd win the bet was at least 0.5. But this was too hasty. Perhaps you have only £0.50, and you really need £0.80 to pay for a bus fare home, and consider amounts below that of virtually no use to you. Then you might be willing to pay out your £0.50 for the bet even if your credence that you would win the bet is well below 0.5, just on the off chance that you might get the bus money you need. Thus even where a person does engage in explicit bets over money, her credences cannot be straightforwardly read off from her betting behaviour. What is needed is some measure of the value that she places on each outcome: that is, the 'utility' of each outcome. Exactly what utilities are and how they might be elicited is debatable, but the hope is that the overall pattern of an agent's actual and dispositional choice behaviour either fixes or elicits both credences and utilities simultaneously. We will return to this point when discussing the 'representation theorem' in section 3.5.

I've now given a brief outline of the probability framework (section 3.2) and introduced credences (section 3.3). In the next section I bring these two things together.

3.4 The credence framework

A probability space consists of a set of states (Ω), an algebra ($\mathcal{F}$), and a probability function. We can use such a space to represent an agent's epistemic state: call this interpretation the 'credence framework'. On the credence framework, the probability function is the agent's credence function. The objects of the credence function are the elements of the algebra:

38 THE CREDENCE FRAMEWORK

these must be the propositions that the agent has credences in. And these propositions—being elements of the algebra—are sets of states.[7]

What then are states? The probability framework in itself does not dictate any particular answer to this question. It is our task to figure out how we should interpret states: we can ask, 'What *could* states be, such that the objects of credence are sets of states?' One reading we might try is to take states to be metaphysically possible worlds; on another reading, we might take states to be epistemically possible worlds; or we might take states to be linguistic items, such as sets of sentences. How should we choose between these options? In the next chapter (chapter 4) I will argue for the tenet that credence claims are 'opaque', and thus when deciding what states are—and more generally how the credence framework works—this is a fact that must be taken into account. Chapters 7 and 8 explore some of the ways we might try to accommodate the tenet. For now, I set this question to one side, and turn to consider the other components of the credence framework.

What is the 'algebra'? We know that the elements of the algebra are sets of states, and that each set of states in the algebra must be a proposition—the sort of thing that an agent has credence in. But which propositions exactly should be included in the algebra? Should we include just those propositions that our agent is consciously entertaining? This is unlikely to result in an algebra: an agent might be consciously considering a proposition e_1 and a proposition e_2 without consciously considering their union. A better approach may be to include all those propositions which an agent at a time *can* entertain, in the hope that this will form an algebra: if you *can* entertain e_1 and e_2 then presumably you can entertain $e_1 \cup e_2$. Some issues remain: it seems that the number of propositions that you can entertain will change over time (as you acquire new concepts, for example), and as we shall see the credence framework does not seem to allow for this. But for our purposes here we can go along with this standard answer: when using a probability space to represent an agent's credence function at a time, the algebra will contain all and only the propositions that she can entertain.

---

[7] Here I focus on a common approach to the credence framework due to (Kolmogorov, 1933 (1950))—but there are alternatives. In particular, Bruno de Finetti's approach begins with a set of objects which can be assigned a truth-value: we can think of these as propositions. We then consider the possible ways that truth-values can be assigned to all of these objects: we can think of each such way as a possible world. The set of probability functions is then defined as any assignment of values that lies within what is called the 'convex hull' of these possible worlds (de Finetti, 1931).

THE CREDENCE FRAMEWORK 39

We have already discussed how to interpret the probability function: we interpret it as the agent's credence function, giving her credence in every proposition that she can entertain. A probability function obeys the probability axioms, detailed in section 3.1, and so this obviously places constraints on what an agent's credences can be. Are these constraints plausible? Consider, for example, the second axiom, which states that the probability assigned to Ω is 1. Is it plausible that an agent would assign a credence of 1 to Ω? Many have argued that this is not plausible on the grounds that this requirement is too demanding: this is part of what is known as the problem of 'logical omniscience' (Savage, 1967; Hacking, 1967; Elga and Rayo, Forthcoming). The thought is that there are propositions that hold at every state, but that we might reasonably be less than fully certain of. For example, there are mathematical and logical propositions that are too complex for us to verify, but which are true at every state. Thus (it might be argued) we do *not* always have a credence of 1 in Ω. But this argument requires some unpacking. Which propositions are true at every state will depend on what states are: on some interpretations of states, a true mathematical or logical proposition will be true at every state, but there may be some interpretations of states on which a mathematical or logical proposition may be true at some states and false at others. Thus whether the requirement is too demanding will depend partly on how states are interpreted. And it will also depend on how we understand credence attribution statements. Consider, for example, that we might naturally say that an agent has a credence of 1 in some simple mathematical truth, but a credence of less than 1 in some complex mathematical truth. If both of these mathematical truths hold at all states, then what credence should we say the agent has in Ω? It is not at all obvious. A key task in this book is to explore these sorts of questions, and figure out how the credence framework should be interpreted—and it is only once this work is complete that we can judge whether the probability axioms are too demanding.

A connected debate concerns the status of the credence framework: are the representations supposed to be descriptive or normative? That is, are we trying to represent how agents actually are, or are we trying to represent how agents *should be*? For those who think that we are trying to represent how agents actually are, there is a concern that we are either missing our mark or misrepresenting agents. To return to the problem of logical omniscience described in the last paragraph, if we represent agents as having a credence of 1 in Ω, then the worry is that either we are failing to represent any real agents or we are misrepresenting real agents. There are various

40 THE CREDENCE FRAMEWORK

responses to this point. As discussed in the paragraph above, there may be ways of interpreting the credence framework so that it does not misrepresent in this way. An alternative response is to argue that the representations are *models*, and not every aspect of a model is supposed to represent its target. For example, a small replica of a castle can represent a full-scale castle, and here we wouldn't say that the model is misrepresenting its target by getting the size wrong. Some such response may be open to a theorist who maintains that the aim is to represent real agents.

For those who argue that the representations are normative, the problem of logical omniscience is perhaps less relevant. On this view, we are not trying to represent how agents actually are, but rather to represent how they should be: that is, we are trying to represent ideally rational agents. And (these theorists can argue) an ideally rational agent is logically omniscient in the required sense. We might go on to debate whether even ideal rationality really does require logical omniscience, and (relatedly) what sense of 'rationality' is in use here, and these are important questions for this view. A further question concerns the purpose of the whole enterprise. If the credence framework is designed to represent real agents, then we can see how such representations might be beneficial, for they would help us to predict and explain the epistemic states and choice behaviour of real agents. If on the other hand the credence framework is designed to represent ideal agents, then we will not be able to use the representations straightforwardly to predict and explain the epistemic states and choice behaviour of real agents—though we may be able to use the representations as approximations. How else might the framework be of use, if we take the representations to be normative rather than descriptive? One answer to this question is that we may be able to use the representations as tools to lead us towards a state of rationality: for example, a scientist might turn to the credence framework to calculate the appropriate level of confidence in a given theory. Here I do not attempt to adjudicate whether the representations should be interpreted as descriptive or normative: my aim here has just been to describe in broad outline the positions available.

In the next section I turn to a series of arguments in favour of the claim that the epistemic state of rational agents can be represented by a credence function that obeys the probability axioms.

3.5 Arguments for the probability axioms

As we have seen, there are reasons to question whether the epistemic states of real agents can be realistically represented using a credence function that obeys the probability axioms. And we might question whether even an ideally rational agent's credence function will obey these axioms. However, there are several arguments in favour of this latter claim in the literature, and below I briefly outline three such arguments: the dutch-book argument, the accuracy argument, and the representation theorem.

I begin by describing the dutch-book argument (Ramsey, 1931). A dutch book of bets is a set of bets that is guaranteed to lose you money, and it seems obvious that no rational agent would accept such a set of bets. Yet we can show that for any agent whose credence function violates the probability axioms, there is some dutch book of bets that the agent would accept as fair: that is, the agent is 'dutch-bookable'. To see an example of this, take an agent who has a credence of 0.5 in Ω—and so violates the second axiom. We can then offer this agent the following bet: she gets £0.5, and pays out £1 iff Ω. This bet is fair by the agent's own lights, and yet it is guaranteed to lose the agent money, and so this agent is dutch-bookable. An agent who is dutch-bookable is irrational—for what rational agent would accept as fair a set of bets that is guaranteed to result in a loss?—and so we have shown that an agent who violates the second probability axiom is irrational. We can produce similar arguments to show that any agent who violates any of the probability axioms is irrational, and thus show that any rational agent's credence function is a probability function.

A number of objections have been raised against dutch-book arguments. For a start, is it really the case that an agent who is dutch-bookable in the sense described is irrational? If there were unscrupulous bookies roaming about offering these bets, then perhaps an agent who accepted them would be at a practical disadvantage—but firstly it's not obvious that there are bookies who operate in quite this way, and secondly merely showing that an agent is at a practical disadvantage is not the same as showing that an agent is irrational in an epistemic sense. The dutch-book argument has been refined in response to these and other objections (Christensen, 1996; Hájek, 2005b; Hájek, 2008). A further concern is that it is not immediately obvious what is meant by an agent being *guaranteed* to lose money. To get an argument for probabilism, we need to take this to mean that the agent would lose money *at every state*, but as we have seen, there are many possible interpretations of states. Will there correspondingly be many possible interpretations of dutch-bookability? Which of these notions of dutch-bookability is connected with irrationality and why? Without a compelling answer to this question, the ambiguity in the notion of dutch-bookability casts doubt on the idea that there is a tight relationship between dutch-bookability and irrationality (Mahtani, 2020).

42 THE CREDENCE FRAMEWORK

The second argument for probabilism is the accuracy argument (Joyce, 1998). To explain this argument, I'll first explain what is meant by 'accuracy' and 'inaccuracy' here. The inaccuracy of an agent's credence in a particular proposition depends on how big the difference is between the agent's credence and the truth-value of the proposition: if a proposition is true then the further the agent's credence is from 1 the less accurate it is; and conversely if a proposition is false then the further the agent's credence is from 0, the less accurate it is. We'll assume some way of measuring this distance that meets various criteria (Joyce 1998), and this gives us a number that represents the inaccuracy of an agent's credence in a particular proposition. Then we get the inaccuracy of an agent's credence function as a whole by summing the inaccuracy of her credence in each proposition. Now let us again take an agent with a credence of 0.5 in Ω, violating the second probability axiom. It is possible to show (though I do not do so here) that there is an alternative credence function which is guaranteed to be more accurate than this agent's credence function. That is to say, the agent's credence function is 'accuracy dominated' by an alternative credence function. This shows that the agent is irrational—for who would have one credence function when another is guaranteed to be more accurate? We can similarly show that any credence function that violates any of the probability axioms is accuracy dominated, and so establish that a rational agent's credence function conforms to the probability axioms. Once again, there are several objections that can be raised against this argument. Firstly, the argument works only if we choose a way of measuring inaccuracy that meets certain criteria—but there is a debate over whether these criteria can be justified (Norton, Forthcoming). Secondly, although accuracy is no doubt a desirable feature of a person's credence function, it is not the only desirable feature, and so it does not automatically follow that an agent with an accuracy-dominated credence function must be irrational (Pettigrew, 2013). And thirdly, once again there is ambiguity around what it means for a credence function to be accuracy dominated: the alternative credence function must be more accurate at every state, but the nature of states is open to interpretation (Mahtani, 2020).

The third argument for probabilism comes from representation theorems (Ramsey, 1931; Savage, 1954). We can begin with the idea of a preference relation for an agent at a time. If we take two alternatives, O_1 and O_2, then an agent might like O_1 at least as much as O_2 ($O \geqslant O_1$) and/or like O_2 at least as much as O_1 ($O_2 \geqslant O_1$). Intuitively, there are certain constraints on the preferences that a rational agent can have. For example, a rational agent who likes O_2 at least as much as O_1, and also likes O_3 at least as much as O_2, must like O_3 at least as much as O_1. We can call such constraints the 'axioms of preference'.[8] Now we suppose that an agent can have preferences not just over what we would think of as outcomes, but also over lotteries on those outcomes. For example, an agent might prefer a lottery that pays out £10 iff P and £2 otherwise, over a lottery that pays out £5 iff P and £2 otherwise—thereby showing she prefers £10 to £5. Thus if we were to attempt to represent this agent's preferences by giving a 'utility function'—assigning a number (the utility) to each alternative—then we should assign a higher number to £10 than to £5. Let's also suppose that this agent prefers a lottery that pays out £10 iff P and £5 otherwise, over a lottery that pays out £10 iff Q and £5 otherwise—thereby showing (given that we've established that this agent prefers £10 to £5) that she has a higher credence in P than she has in Q. Thus if we were to use numbers to represent this agent's epistemic attitude towards the relevant propositions—that is, if we were to attempt to give her credence function—then we should assign a higher number to P than to Q. We can see, then, how an agent's preferences over these lotteries constrains the utility and credence functions that we should use to represent her preferences and epistemic attitudes. The proponents of representation theorems show (though I don't attempt to reproduce their arguments here) that—given certain assumptions—an agent's preferences over these lotteries constrain the credence and utility functions that can be used to represent them almost to the point of uniqueness.[9] And the functions that result display certain features: in particular the credence function conforms to the probability axioms. Thus we have a proof that a rational agent's credence function obeys the probability axioms: a rational agent's preferences conform to the axioms of preference, and the epistemic state of such an agent can be represented by a credence function that conforms to the probability axioms. There are a range of versions of this sort of argument,[10] and various objections to be found in the literature—many focusing on whether rational

---

[8] Arguably these axioms themselves require justification, and something like a dutch-book argument or 'money pump argument' is needed here—in which case we should not really see representation theorems as an entirely distinct sort of argument from dutch-book arguments. Here for the purposes of this introductory chapter I follow much of the literature in introducing these as separate arguments.

[9] The exact details vary between accounts, but typically the result is a fully unique credence function, and a utility function that is unique up to positive linear transformation: for example, if a utility function that assigns x to O_1 and y to O_2 is permitted, then so is a utility function that assigns $ax + b$ to O_1 and $ay + b$ to O_2 (for any b and any positive a).

[10] In particular, Joyce constructs a very general representation theorem in (Joyce, 1999).

44 THE CREDENCE FRAMEWORK

agents are required to conform to the axioms of preference, and whether the further assumptions made in the course of the argument are warranted.

Here I have briefly outlined three arguments designed to show that rational agents, at any rate, have epistemic states that can be represented using the credence framework. I turn now to describe some further rules of rationality that have been proposed.

3.6 Other rules of rationality

There are two different sorts of rules of rationality relevant to the credence framework: synchronic rules (which constrain an agent's credence function at a particular point in time) and diachronic rules (which constrain how an agent's credence function can rationally change as time passes). We have already seen a synchronic rule of rationality: the rule that a rational agent's credence function at a time conforms to the probability axioms. I turn now to a diachronic rule—the principle of conditionalization. This is a widely (but not universally) accepted principle of the credence framework. This principle concerns how an agent's credence function may rationally change over time. Let us say that at a time t_0, an agent's credence function is given by Cr_0. And let us suppose that between t_0 and later time t_1, the agent learns proposition e (and gains no other evidence). What then should the agent's credence function Cr_1 at t_1 be? Obviously both Cr_0 and Cr_1 are rationally required to conform to the probability axioms—but are there any further constraints? The principle of conditionalization does impose a further constraint: Cr_1 must be Cr_0 'conditionalized on e'. To see what this means, I first need to explain conditional probability.

Take any two propositions e_1 and e_2 assigned values by a probability function P. The probability function will also assign a value to their intersection $e_1 \cap e_2$.[11] We can then define the probability of e_1 *conditional on* e_2—written $P(e_1|e_2)$—as $P(e_1 \cap e_2)/P(e_2)$. Intuitively this is the probability of e_1 *given* e_2: how likely is it that e_1 (together with e_2) obtains, given that e_2 obtains? This value is undefined where $P(e_2)$ is zero. There are those who take the formula $P(e_1|e_2) = P(e_1 \cap e_2)/P(e_2)$ to be a way of defining

---

[11] The probability function will assign values to every element of an algebra. If it assigns a value to e_1 and a value to e_2, then it must assign a value to e_1^C and e_2^C, and so to the union $e_1^C \cup e_2^C$, and so to the complement of this union $(e_1^C \cup e_2^C)^C$, which is equivalent to the intersection $e_1 \cap e_2$.

OTHER RULES OF RATIONALITY 45

conditional probability; and there are those who think that the notion of conditional probability is intuitively clear without this definition—and that in some cases the definition may not apply (Hájek, 2003). I do not delve into this issue, but for the purpose of this book I assume that the equivalence holds in all the cases that we consider.

Having introduced conditional probability, I can now explain what the principle of conditionalization requires. According to this principle, when a rational agent learns that e, her new credence in any proposition e_x is equal to her old credence in e_x conditional on e. Thus where our agent learns e (and only e) between t_0 and t_1, then for any proposition e_x, $Cr_1(e_x) = Cr_0(e_x|e)$. Note that it follows from this principle that the agent's new credence in e is 1: $Cr_1(e) = Cr_0(e|e) = Cr_0(e \cap e)/Cr_0(e) = Cr_0(e)/Cr_0(e) = 1$. Thus on learning that e, our agent (if rational) becomes certain that e, and her new credence in every proposition is equal to her old credence in that proposition conditional on e. According to the principle of conditionalization, this is the only way that a rational agent's credence will change as time passes: she will never forget any information; she will never change her credences without evidence; and when she does gain evidence, her credences will change by conditionalization and in no other way.

Conditionalization may strike you as intuitively the right way for an agent to change (or 'update') her credences. And—though the status of some of these diachronic arguments has been disputed—there are dutch-book and accuracy arguments in favour of the principle of conditionalization (Teller, 1973; Brown, 1976; Lewis, 1999; Greaves and Wallace, 2006). The principle raises some questions, however. Is it really the case that a rational agent would never change her credences except by conditionalization? Can't an agent be rational and sometimes forget something? Can't an agent figure something out just by thinking about it, and so rationally change her credences without gaining any new evidence? Can't an agent become aware of a new possibility, and so expand the set of propositions that she has credences over—a move that the principle of conditionalization leaves no room for? And what are we to do about indexical claims—such as the claim that it is currently 2 p.m.: it seems that the standard principle of conditionalization does not accommodate our changes in attitude towards these sorts of claims. Finally, is it really the case that every case of learning involves becoming *certain* of some proposition?

In response to these issues, theorists have proposed a number of variations on the principle of conditionalization. Firstly, many theorists reject the standard principle of conditionalization in favour of 'Jeffrey

Conditionalization' (Jeffrey, 1965 (1990)). The underlying idea is that we can drop the (perhaps implausible) assumption that learning always results in assigning a credence of 1 to new evidence gained, and yet still retain the essential thought that a rational agent's credence function after a learning event is constrained by her earlier credence function. Secondly, there are those who claim that conditionalization is not the only way in which a rational agent can update her epistemic state. In some cases of 'awareness growth', the algebra can expand as an agent conceives of propositions that she had not considered before. Analogues of the principle of conditionalization have been proposed to constrain the way that awareness growth takes place for a rational agent (Karni and Vierø, 2013; Bradley, 2017). And thirdly, there are theorists who supplement the principle of conditionalization so as to accommodate indexical claims (Titelbaum, 2016). In general for the purposes of this book, we will deal only with the standard principle of conditionalization, but the claims that I make will also apply to these variations on the standard principle.

Besides the principle of conditionalization, various other rules of rationality have been proposed—though these are more controversial. Many of the other proposed rules of rationality are principles of 'deference'. A deference rule requires you to treat some probability function as expert, and so defer to it in the following sense: you will adopt any credences, conditional on the claim that those are the probabilities assigned by the expert probability function. From a principle of deference, together with conditionalization, it follows that if you were to learn that an expert has a particular credence in a particular proposition, then (if rational) you would adopt that same credence in that proposition. Some argue that you ought to treat your future self as an expert and so defer to that future self: this is the 'Reflection Principle' (van Fraassen, 1984). Some argue that you ought to treat the objective chance function as an expert and so defer to that chance function: this is the 'Principal Principle' (Lewis, 1987). And there are various other principles of deference, along with principles of disagreement, which concern peers rather than experts. I will return to some of these principles in chapter 5.

I turn now to one final restriction that we might think rationally governs an agent's credences. This is a synchronic, rather than diachronic, restriction. We have seen that there is already one sort of synchronic restriction on an agent's credences: a rational agent's credence function conforms to the probability axioms. And some of the principles of deference mentioned in the last paragraph are also synchronic restrictions—albeit more

CHAPTER SUMMARY   47

controversial restrictions. But are these all the restrictions? Provided that I conform to these, then can I—rationally—have any credences that I like? Suppose that you and I have exactly the same evidence (e), which includes observing a long series of coin tosses with an even mix of heads and tails. You have a credence of ½ that (HEADS) the next coin toss will land heads— and let us suppose that your credences conform to the probability axioms and to the other principles of rationality described above. My credences also conform to the probability axioms and the other principles above, but I have a credence of 1/10 that the next coin toss will land heads. Can we both be rational? It might be thought that one of us must be violating the principle of conditionalization: if we have gained the same evidence, and both conditionalized, then why haven't we ended up with the same credences? The answer is that we might not have *started* with the same credences. The principle of conditionalization requires that your new credence function is your old credence function conditionalized on whatever evidence you have gained. If we started out with different credence functions, then even though we've gained the same evidence e and both conditionalized, we could end up with different credence functions. It might be that you started out with a credence function that assigned ½ to HEADS conditional on e, whereas I started out with a credence function that assigned 1/10 to HEADS conditional on e. We can call the credence functions that we started out with our *priors*. Both of our priors, let's assume, conform to all of the rules of rationality described so far. A true subjective Bayesian would insist that both of us are rational: the rules of rationality are as stated above, and no further constraints are placed on an agent's credences. Other Bayesians introduce further rules of rationality—some going so far as to rule that there is only one rational prior credence function (Hedden, 2015). On this view, any two agents with the same evidence must have the same credences, on pain of irrationality. I don't take a stand on this here, but just note the debate.

3.7 Chapter summary

In this chapter I have given an overview of the Bayesian framework. I began in section 3.2 by describing the probability framework, with its abstract mathematical 'probability spaces'. I showed that there are a variety of ways of interpreting these spaces, and clarified that we will be focusing on using a probability space to represent an agent's epistemic state by giving her credences. Then in section 3.3 I discussed the nature of credences and

48 THE CREDENCE FRAMEWORK

their relation to betting behaviour and choice behaviour more generally. In section 3.4 I turned to the probability axioms and considered whether these are plausible when we interpret a probability space as giving an agent's credences. In section 3.5 I described some of the arguments for the probability axioms, and then in section 3.6 I surveyed a range of other rules of rationality that are sometimes put forward. This completed the introduction to the credence framework.

This credence framework gives us a way of representing an agent's epistemic state. The framework is used in many different fields: it is used, for example, in science, in statistics, in economics, and in policy choice. Indeed, you can find instances of its use in almost every discipline. It is therefore very important to understand the framework and interpret it correctly. In the next section I turn to argue for the central tenet of this book: that credence claims—that is, claims attributing credences to agents— are opaque.

4
Credence Claims Are Opaque

4.1 Introduction

As we saw in chapter 3, on the credence framework a person's epistemic state is represented by a function from propositions—or 'events'—to numbers. And as we saw in chapter 2, there is a lot of debate and disagreement over the nature of propositions. I turn now to bring these two areas together.

A central tenet that I put forward in this book is that credence attribution statements create an opaque context: often I will condense this tenet into something a bit pithier by saying that credence claims are opaque. In this chapter I argue for this central tenet. Then in chapters 5 and 6, I trace some of the implications of the tenet, and in chapters 7 and 8 I consider how we might interpret the credence framework to accommodate the tenet.

4.2 The 'it's just obvious' argument

I begin by explaining what it would mean to say that credence claims are opaque, and for this I return to Frege, first introduced in chapter 2. On Frege's account the referent of a name is the object to which it refers, the referent of a predicate is a function from objects to truth-values, and the referent of a sentence is a truth-value. For Frege, the referent of a sentence (i.e. its truth-value) is determined by the referents of its parts. We saw (in section 2.3) that this account led to a puzzle, which we can illustrate with the following two sentences:

(4a)   Tom believes that George Orwell is a writer.
(4b)   Tom believes that Eric Blair is a writer.

These two sentences are identical, except for the fact that the name 'George Orwell' in (4a) is replaced in (4b) by the name 'Eric Blair'. Let us suppose that these sentences are uttered in the same context, so that each element of

---

The Objects of Credence. Anna Mahtani, Oxford University Press. © Anna Mahtani 2024.
DOI: 10.1093/oso/9780198847892.003.0004

50 CREDENCE CLAIMS ARE OPAQUE

the two sentences has the same referent. For example, 'Tom' in (4a) refers to the very same person that 'Tom' refers to in (4b). And of course, though 'George Orwell' and 'Eric Blair' are different names, they have the same referent because they both refer to the very same person. This leaves Frege with a puzzle: if each part of sentence (4a) has the same referent as the analogous part of sentence (4b), then—given that the referent of a sentence is determined by the referents of its parts—sentences (4a) and (4b) ought to have the same referent. And yet sentences (4a) and (4b) can have different referents, for the referent of a sentence is its truth-value, and (4a) and (4b) can have different truth-values. As we saw, Frege responded to this problem by introducing his account of sense.

Let's focus in here on the phenomenon that created this puzzle for Frege. The problem is that whether it is true to say that a subject believes that some object has some feature can depend on how that object is designated. To explore this point further, I begin by recalling from chapter 2 that some designators are rigid and some are non-rigid. Names are generally taken to be rigid designators: a name refers to the same object at every metaphysically possible world. As a result any two names with the same extension will also have the same intension. In contrast, definite descriptions have a non-rigid reading: a definite description can refer to different objects at different metaphysically possible worlds. Two definite descriptions with the same extension might have different intensions, and similarly a name and a definite description can have the same extension but different intensions. In some contexts (extensional contexts) a designator can be substituted for any other designator with the same extension without affecting the truth-value of the sentence. To see an example of this, consider the sentence 'George Orwell is a writer'. We can safely substitute the name 'George Orwell' for any other name with the same extension ('Eric Blair', or 'the winner of the 1946 Hugo Prize', for example) and the truth-value of the sentence will stay the same. In some contexts (intensional contexts) it is not the case that we can safely substitute any designator with the same extension, but we can safely substitute any designator with the same intension. Certain modal contexts work like this, and we will see examples of this sort of context in chapter 5. But in some contexts (hyperintensional contexts) it is not the case that we can safely substitute any designator with the same extension, nor is it the case that we can safely substitute any designator with the same intension. In hyperintensional contexts, switching one name for a co-referring name may change the truth-value of the sentence—even though the two names have the same intension.

THE 'IT'S JUST OBVIOUS' ARGUMENT   51

We can see from sentences (4a) and (4b) that belief attribution claims are hyperintensional contexts: substituting 'Eric Blair' for 'George Orwell' can change the truth-value of the sentence, even though 'Eric Blair' and 'George Orwell' are co-referring proper names and so have the same intension.

Expressions such as 'intensional' and 'hyperintensional' are most at home in discussions of possible worlds semantics, but in various places throughout the book it is important to be able to draw a distinction between contexts that are intensional and contexts that are hyperintensional, and in these places I will use these terms. Often though the distinction between intensionality and hyperintensionality will be less important, and here I will use the expression 'opaque' to refer to a context that is hyperintensional: that is, a context in which we cannot safely substitute co-referring designators, whether they are names or definite descriptions. I like the term 'opaque' because I think that it captures the underlying idea well. To say that belief attribution claims are opaque implies that agents don't have beliefs about objects *directly*, but rather through the medium of some sort of designator. Tom's belief is not about George Orwell the person, as such, but rather about George Orwell under a particular sort of way of thinking about him: it seems that must be so, because Tom doesn't have the same belief about Eric Blair— and yet Eric Blair is the same person as George Orwell. Thus when designators appear within the context of a belief claim, their role is not just to pick out an object: they are not transparent designators that simply show us the object itself, but rather opaque designators that present us with an impression of the object from a particular angle, rather like the shadows in the image on the cover of this book. And there is something very interesting (and perhaps troubling) about this idea, for it seems to imply that there is a gulf between the world and our representations of it: we do not have beliefs about objects in themselves, but only through some sort of mediation. However, I do not delve further into these metaphysical worries in this book, but focus on the implications that concern users of the credence framework.

The problem that Frege faced, then, is that belief claims are opaque. And the aim of this chapter is to argue that an analogous claim holds for credence claims. That is, when we say that an agent has some particular credence that some object has some feature, it matters how that object is designated.[1] The

---

[1] It can also matter how the feature is described: within the context of a belief/credence attribution statement, we similarly cannot safely substitute a predicate for another predicate with the same extension—or even the same intension. I focus on the designation of objects for simplicity.

52 CREDENCE CLAIMS ARE OPAQUE

first reason to think so is that it seems obvious that Frege's puzzle arises for credence attribution statements just as it does for belief attribution statements. For example, statement (4d) does not seem to follow from statement (4c). That is, (4c) can be true while (4d) is false:

(4c)   Tom has a credence of 0.8 that George Orwell is a writer.

(4d)   Tom has a credence of 0.8 that Eric Blair is a writer.

If that's right—that is, if (4c) can be true while (4d) is false—then credence claims are opaque. This is the key tenet I want to argue for in this chapter. It may seem pretty obvious, but the implications are wide-reaching, so it's important that I take the time to really convince you. Later in this chapter (sections 4.6–4.8) I consider and reject a theory on which belief claims and credence claims are not opaque in quite the way that I claim: on this theory, (4c) and (4d) must be true or false together. But before turning to this alternative theory, I give three sorts of reasons to accept that credence claims are opaque: reasons relating to choice behaviour (section 4.3), reasons connected with omniscience (section 4.4), and reasons relating to conditionalization (section 4.5).

4.3 Choice behaviour

Users of the credence framework typically assume that there is a close relationship between an agent's credence function and her choice behaviour. As discussed in chapter 3, there are various views about what sort of choice behaviour is relevant, and about what exactly the relationship is supposed to be between the relevant choice behaviour and an agent's credence function. On some accounts, the relevant choice behaviour is just betting behaviour, whereas others take a broader view as to the types of choice behaviour that are relevant. And for some, there is a tight logical relation between an agent's credence function and the relevant actual or dispositional behaviour, while for others the relationship is looser, and in some cases normative rather than descriptive. Below I look at a few accounts within this range, and argue in each case for the tenet that credence claims are opaque.

Let us begin with a view on which there is a tight logical relation between an agent's credence function and her dispositional betting behaviour. On this view, an agent's credence function *just is* her disposition to accept or

CHOICE BEHAVIOUR   53

reject certain bets. Thus an agent with a credence of 0.5 in HEADS will accept any bet that costs less than £5 and returns £10 iff HEADS, and reject any bet that costs more than £5 and returns £10 iff HEADS: any agent who did not have this dispositional attitude towards the relevant bets would not count as having a credence of 0.5 in HEADS. Donald Gillies (Gillies, 2000, pp. 55–8) describes such a view, which is related to the approach of the early de Finetti (de Finetti, 1931). This view has some virtues, in that it makes clear what credences are, and how to decisively test for them. But it also faces some problems, described in chapter 3: agents may not value money linearly, and money may not be the only thing that agents value.[2] Let's set these problems aside for now, and temporarily accept that an agent's credence function stands in a tight logical relationship with her betting behaviour. Does it follow that credence claims are opaque?

Let's start with the obvious point that a seemingly rational agent—Tom, for example—might bet differently over the two claims (4e) and (4f) below:

(4e)   George Orwell is a writer.

(4f)   Eric Blair is a writer.

For example, Tom might accept any bet which costs less than £9 and pays out £10 iff (4e) is true, whereas he might reject all such bets on (4f) unless they cost less than £2. On the view that we are considering, an agent's credence in a proposition is logically related to her betting behaviour over that same proposition. Tom's betting behaviour over (4e) and (4f) is different, and so we seem forced on this view to conclude that a rational agent can have different credences in (4e) and (4f)—and so that credence claims are opaque. If we assume further that the objects of credence are propositions, it seems to follow that (4e) and (4f) must express different propositions.

But this is too quick. Does the fact that Tom might bet differently over (4e) and (4f) really show that they express different propositions? Consider the following two claims:

(4e)   George Orwell is a writer.

(4g)   George Orwell est un auteur.

---

[2] As discussed in chapter 3, agents sometimes have religious or other reasons why they do not want to bet at all. I count these as cases where money is not the only thing an agent values.

54 CREDENCE CLAIMS ARE OPAQUE

Tom might bet differently over claims (4e) and (4g). To see this, suppose that his grasp of English is a lot better than his grasp of French, and while he understands (4e) perfectly, he does not understand (4g) perfectly because he is unsure what 'auteur' means. Tom might then be willing to accept any bet that costs less than £9 and pays out £10 iff (4e) is true, but unwilling to accept any similar bet involving (4g) unless it costs less than £5. Here Tom's reluctance to bet on (4g) reflects not just his mild uncertainty over George Orwell's profession, but also his uncertainty over what (4g) says about it. Thus Tom might bet differently over claims (4e) and (4g)—but on the standard account of propositions, (4e) and (4g) express the very same proposition, just in a different language. At first glance, then, this looks like a problem for the account on which there is a tight logical relation between an agent's credence function and her betting behaviour: how an agent bets on a claim does not depend just on the proposition that the claim expresses, but also on other features of the claim such as the language used. And the diagnosis of the problem is that the objects of credence are propositions, but bets are waged over sentences.[3] For example, suppose that a bookie offers you a bet verbally. In this case, you are betting on the truth of the sentence uttered. If you are unsure what the sentence means, then in deciding whether to accept the bet you need to consider a range of propositions, and ask yourself for each proposition—how likely is it that the sentence expresses that proposition, and how likely is that proposition to obtain? In deciding whether to accept the bet, you need to consider all of this. For when the reckoning comes, whether you win or lose the bet will depend on whether the proposition that the sentence actually expressed was true.[4]

How could we respond to this problem, if we wished to maintain the view that there is a tight, logical relationship between credences and betting

---

[3] More accurately, bets are waged over *utterances*. To see this, suppose I were to write 'this horse will win' on a placard, place it in front of a particular horse, and offer you a bet over what I have written. You might be prepared to bet highly. If I then move the placard (on which the token sentence is written) and place it in front of a different horse—or perhaps even in front of the very same horse but from a different and unrecognizable angle—then you may bet differently. Many thanks to Braun for the shape of this example.

[4] In some real-life cases (imagine, for example, betting with a child), whether you get your payout may depend on how the *bookie* interprets the sentence, rather than on what the sentence actually means. And so you might be wise in your calculations to consider not just what the sentence actually means, but what the bookie thinks it means. But accounts on which credences stand in a tight, logical relationship with betting behaviour are plausible only if we abstract away from such real-life cases of betting: after all, in real-life cases, the honesty and financial stability of the bookie should also figure in your calculations.

behaviour? One possibility is to claim that the objects of credence are not propositions (as they are generally understood), but rather sentences. This claim faces the difficulties mentioned in chapter 2 for an analogous claim about belief: the objects of belief cannot be token sentences, for sometimes there is no relevant token sentence available to be the object of a belief; and the objects of belief cannot be sentence types, because two beliefs with different truth-values can correspond to the same sentence type. How else might we respond to the problem? Another option is to claim that your credence in a proposition is fixed by how you are disposed to bet over any token sentence that has two features: it expresses the relevant proposition, and it is a sentence *that you fully understand*.[5] In this case, Tom's credence that George Orwell is a writer is fixed by his betting behaviour concerning sentence (4e) which he understands, and not sentence (4g), which he does not. This seems right.[6]

If we turn back, then, to the two sentences (4e) ('George Orwell is a writer') and (4f) ('Eric Blair is a writer'), it seems that we have two options. The first is to say that Tom understands them both. Given that Tom would bet differently over them (he would bet more that George Orwell is a writer than that Eric Blair is a writer), it follows that his credence in each must be different. From this it follows that credence claims are opaque. This is my preferred response. Alternatively, we might say that Tom does not understand both of (4e) and (4f). This leaves open the option of saying that credence claims are not opaque: Tom's credence that George Orwell is a writer is the same as his credence that Eric Blair is a writer—it's just that his credence is not accurately reflected by his betting behaviour over at least one of the relevant sentences, because at least one of the relevant sentences is not fully understood. Perhaps we could plausibly argue that Tom doesn't understand, say, (4f), if he has just picked up the name 'Eric Blair' from his friends, and is using it with a quite mistaken idea about who Eric Blair is. But we can fill out our example by imagining that Eric Blair is a regular visitor to the café where Tom works, and so Tom cannot plausibly be said not to know who Eric Blair is. Nor can he plausibly be said not to know who George Orwell is: Tom has read lots of George Orwell's books, and we can suppose

---

[5] You need to both grasp the meaning of the sentence and also be aware of any relevant features of the context.

[6] Another appealing option is to follow Donald Davidson in emphasizing the role that 'theories of meaning' play in determining choice behaviour—where a theory of meaning gives an interpretation of various utterances (Davidson, 1973). For simplicity I do not pursue this option here.

56 CREDENCE CLAIMS ARE OPAQUE

even attended a book signing where he saw George Orwell—albeit looking unrecognizable from the regular café visitor. It seems, then, that Tom understands both (4e) and (4f). Given that he bets differently over each, it follows—on the view that there is a tight logical relationship between credences and betting behaviour—that Tom has a different credence in (4e) and (4f), and so that credence claims are opaque.[7]

Let us turn now to alternative accounts which focus on the relationship between an agent's credence and utility functions, and her more general choice behaviour—as opposed to specifically her betting behaviour. On some views the relationship here is a tight, logical relationship, so that an agent will—as a matter of logical necessity—always choose in line with her credence and utility functions, for an agent's credence and utility functions are simply a representation of her actual and dispositional choice behaviour. On other views, the relationship may be looser, so that the laws relating an agent's credence and utility functions to her choice behaviour are psychological rather than logical laws. And on other views, the laws are normative rather than descriptive, so an agent's credence and utility functions determine what it is *rational* for her to choose, but not what she will actually choose. On all these views, I argue, we have reason to hold that credence claims are opaque.

To see this, let us return to our case of Tom, who works in a café with a regular customer known as 'Eric Blair', and in his spare time reads the books of his literary hero 'George Orwell', without realizing that these names refer to one and the same person. Let us suppose that while he is walking over to take an order from his regular customer Eric Blair, he sees a report in the newspaper that states that George Orwell will be visiting a particular bookshop on the other side of the city that day. Tom shouts an apology to his customer and rushes out in the direction of the bookshop. Why does Tom head towards the bookshop, while shouting an apology? If we want to explain or justify Tom's behaviour in terms of his beliefs and desires, then a key part of the explanation will be that Tom believes that George Orwell is in the bookshop (which is why he is rushing towards it), and also believes that Eric Blair is in the café (which is why he is shouting an apology). If belief

---

[7] Perhaps there is room here to argue that even with all of his information, there is still a sense in which Tom does not understand both (4e) and (4f), because he does not recognize that they are the same proposition. But even if this position can be spelled out, the claim that there is a tight and logical relationship between credences and betting behaviour would then lose its appeal: the relationship would hold only when the sentence over which the bet is waged is *understood*—where the level of understanding required is unrealistic.

claims were not opaque in the way I've suggested, then the fact that Tom believes that George Orwell is in the bookshop would automatically entail that he believes that Eric Blair is in the bookshop—in which case we would be stuck for an explanation of his apology. And the fact that Tom believes that Eric Blair is in the café would automatically entail that he believes that George Orwell is in the café—in which case we would be stuck for an explanation of his movements. A charitable interpretation of Tom's behaviour is available only if belief claims (and desire claims) are opaque.

Similarly, Tom's behaviour can be seen to cohere with his credences (and utilities) only if credence claims (and utility claims) are opaque. This holds whether the relation between Tom's behaviour and his credence and utility functions is taken to be tight and logical or somewhat looser, and whether it is taken to be descriptive or normative. To see this, let us suppose for the sake of argument that credence claims are not opaque. Then whatever Tom's credence that Eric Blair is in the café, Tom must have the same credence that George Orwell is in the café. But what credence could Tom have here that fits with his behaviour—where his behaviour involves rushing away from the café towards the bookshop while simultaneously shouting an apology? If Tom has a high credence that Eric Blair—and so, George Orwell—is in the café, then why does he rush away from it? And if Tom has a low credence that Eric Blair is in the café, then why does he shout an apology? If credence claims are not opaque, then it is hard to see how credences (together with utilities) could be used to give a causal explanation for Tom's behaviour. Requiring that the relationship between credences and utilities and behaviour is tighter does not make this problem any easier, and neither does taking the relation to be normative rather than descriptive, for it is no easier to *justify* Tom's behaviour using non-opaque credence claims than it is to causally explain or predict it, and it would be uncharitable to categorize Tom's behaviour as simply unjustifiable. Thus for a range of relations that are taken to hold between an agent's choice behaviour and her credence and utility functions, these relations hold only if credence (and utility) claims are opaque.

4.4 Omniscience

People sometimes complain that the credence framework places unreasonable demands on an agent. On the credence framework, we represent an agent as having a credence function that obeys the probability axioms, and these axioms require a credence of 1 in any tautology. As we have seen,

58 CREDENCE CLAIMS ARE OPAQUE

precisely what counts as a tautology will depend on the details of the framework, but on many of the ways of filling in these details, truths of logic, at any rate, will count as tautologies. This has led many to object that the probability axioms are too demanding: some logical truths are far from obvious, and even the best logician will not have a credence of 1 in all of them. This is part of what is called the problem of 'logical omniscience'. As discussed in chapter 3, defenders of the credence framework sometimes respond to this problem by claiming that the rationality requirements should be understood as requirements for *ideal* rationality, and that there is nothing counterintuitive about claiming that none of us is ideally rational. Relatedly, some argue that the credence framework gives us *models* rather than complete descriptions of agents, and these are to be understood as idealizations. Whether these responses are adequate is open to debate, and I do not attempt to settle this issue here, for—unless we accept that credence claims are opaque—there are much larger issues to address.

If credence claims are not opaque, then the credence framework would require of an agent not just logical omniscience, but something akin to straightforward omniscience. To see this consider first that a rational agent is presumably required to be certain of the following logical truth if she can entertain it:

(4h)   If George Orwell exists, then George Orwell is George Orwell.

If credence claims are not opaque, then it follows that (4h) expresses the same object of credence as (4i):

(4i)   If George Orwell exists, then George Orwell is Eric Blair.

On this view, any rational agent will be certain of (4i) if she is able to entertain it. But this doesn't seem right, for an agent can be perfectly rational—ideally rational even—without being certain of this fact. Intuitively, to be certain of (4i) doesn't just require extraordinary reasoning skills, which might, plausibly, be what is required for ideal rationality, but also requires knowledge of the world, and it seems that an agent can be perfectly rational while lacking such knowledge. An ideally rational agent might be required to know all a priori truths, but not, surely, all a posteriori truths. There doesn't seem to be any good sense of 'rational' such that certainty of (4i) is required for ideal rationality. This problem is widespread: the credence framework, combined with the assumption that credence

OMNISCIENCE   59

claims are not opaque, requires that a rational agent is certain of any true identity statement with a form similar to (4i) that she can entertain.[8]

Chalmers (Chalmers, 2011a, p. 599) discusses the implications of insisting on a similar rationality requirement—the requirement that a rational agent should have credence 1 in all (metaphysically necessary) identity claims. Chalmers argues that this would entail that a rational agent is required to be omniscient. To grasp Chalmers's argument for this claim, let us take some particular true statement, A. We can easily coin a name for the truth-value of claim A: let's follow Chalmers and say that the truth-value of A is hereby named 'Fred'. As it happens, because claim A is true, 'Fred' and 'True' co-refer, because they both name the very same truth-value. And because 'Fred' and 'True' are both names, they are rigid designators, and so refer to the same object at every possible world. Thus 'Fred = True' is a necessarily true identity statement, and so (under the rationality requirement that Chalmers is discussing) any rational agent would be required to have credence 1 in this claim. From the claim that 'Fred' names the truth-value of statement A, together with the claim that Fred is identical to the truth-value True, a rational agent can infer that A is true. This amounts (Chalmers argues) to the claim that a rational agent is required to have credence 1 in every true claim: in other words, omniscience—not just logical omniscience, but straightforward *omniscience*—is now a requirement of rationality. This is clearly an unwelcome conclusion—partly because (as Chalmers points out) it undercuts the motivation for Bayesian epistemology, which is to model uncertainty.

Can we use Chalmers's argument to prove that credence claims must be opaque, on the grounds that otherwise all rational agents are omniscient? Take a true statement for which we have coined a name ('Fred') for its truth-value, and an agent who is a competent user of that name, and so knows (has a credence of 1, let's suppose) that 'Fred' names the truth-value of this state-ment. This agent has a credence of 1 that Fred is Fred. If credence claims are not opaque, then it follows (given that 'Fred' and 'True' name the same truth-value) that this agent also has a credence of 1 that Fred is True. And so it

---

[8] A theorist who applies guise Russellianism to credences may have a response to this line of thought. For such a theorist might claim that (4i) is actually a priori, in which case (such a theorist might argue) any rational agent ought to be certain of it. The theorist does then admittedly seem to be led to the conclusion that all rational agents are omniscient, but (this theorist can claim) rational agents are not required to be omniscient in any troubling or surprising sense: rational agents are indeed required to be certain of all true identity claims, but not of all true identity claims under every guise. I discuss guise Russellianism in more depth later in this chapter.

60 CREDENCE CLAIMS ARE OPAQUE

follows (given that the agent is rational) that the agent has a credence of 1 in the relevant statement. It follows, then, that for every true statement such that a name for its truth-value has been coined, and for any rational agent who is a competent user of that name, that agent will have a credence of 1 in that statement. Of course, we have not in fact coined names for every true statement, so the claim that all rational agents are omniscient does not follow directly—but a conclusion to more or less the same effect can be established. Faced with any statement, if an agent coins a name for its truth-value, then if in fact that statement is true, rationality will require her to be certain of that statement. This in itself is a very unwelcome result.

The way to avoid this result is to claim that credence claims are opaque. Then we can simply state that an agent might be required to be certain that Fred is Fred without being certain that Fred is True; and similarly, an agent can be required to be certain of (4h) without being required to be certain of (4i). This gives us another reason to think that credence claims are opaque.

4.5 Conditionalization

Chalmers (Chalmers, 2011a) has argued against 'referentialism' about credence. Referentialism about credence is closely related to the idea that credence claims are not opaque. A referentialist about credence would hold that if a and b are two names for the same object, then an agent's credence that a is F must be the same as the agent's credence that b is F, on the grounds that (on this view) 'a is F' and that 'b is F' express the same proposition. Chalmers's argument against referentialism about credence is part of a wider project: from the claim that referentialism about credence is false, Chalmers infers that referentialism about belief is false, by arguing that the objects of belief and credence are one and the same. Here I do not consider this wider argument, but focus on Chalmers's argument for the claim that referentialism about credence is false.

Chalmers's argument focuses on an example in which Olivia is a clinical researcher, interested in a particular disease ('dreadfulitis' (Braun, 2016)) and two possibly related genes (the alpha gene and the beta gene). At the start of the day, at t_0, Olivia sees from her patient list that she will be getting some information about Dr Jekyll at t_1 and some information about Mr Hyde at t_2—and she does not realize at this point that Dr Jekyll is Mr Hyde. Let JA/HA, JB/HB, JD/HD be the propositions that Jekyll/Hyde have alpha, beta, and dreadfulitis respectively. At t_0, Olivia has the following credences:

CONDITIONALIZATION   61

$Cr_0(JA) = Cr_0(HA) = Cr_0(JB) = Cr_0(HB) = 0.01$ (because 1% of the population have the alpha gene, and 1% of the population have the beta gene)

$Cr_0(JA|JB) = Cr_0(JB|JA) = Cr_0(HA|HB) = Cr_0(HB|HA) = 0.01$ (because the genes are independent)

$Cr_0(JD|JA) = Cr_0(HD|HA) = 0.1$ (because 10% of people with the alpha gene have dreadfulitis)

$Cr_0(JD|JB) = Cr_0(HD|HB) = 0.2$ (because 20% of people with the beta gene have dreadfulitis)

$Cr_0(JD|JA\&JB) = Cr_0(HD|HA\&HB) = 0.9$ (because 90% of the people with both genes have dreadfulitis)

At t_1, Olivia learns just JA. She conditionalizes on this evidence, and so $Cr_1(JD) = Cr_0(JD|JA) = 0.1$, and $Cr_1(JD|JB) = Cr_0(JD|JA\&JB) = 0.9$. Then at t_2, Olivia learns just HB. What should her credence be at this point in HD? If referentialism holds, then JA is the same proposition as HA, and similarly JB is the same as HB and JD is the same as HD. We have seen that $Cr_1(JD|JB) = 0.9$, and so given referentialism, $Cr_1(HD|HB) = 0.9$, and so if Olivia conditionalizes as she should, $Cr_2(HD) = Cr_1(HD|HB) = 0.9$. But, as Chalmers points out, this is intuitively incorrect: 'referentialism combined with Bayesianism yields a false prediction here' (Chalmers, 2011a, p. 591).[9] Chalmers goes on to give a further argument, but the argument I have outlined here is the one that he claims has the most force (Chalmers, 2011a, p. 593). Chalmers's argument thus gives us a further reason to claim that credence claims are opaque: if we deny that credence claims are opaque, and assume that Olivia is rational and conditionalizes as she should, then it follows that $Cr_2(HD) = 0.9$, and this is counterintuitive.

It seems obvious, then, that credence claims are opaque, and the arguments given in the last three sections further bolster this view. But to fully convince you that credence claims are opaque in the way I have described, I need to thoroughly consider a tempting alternative. I turn to this now.

---

[9] Chalmers claims that if Olivia is rational then $Cr_2(HD) = 0.2$. I do not think this is mandated. On Chalmers's view, as on mine, the claim that Jekyll is Hyde is not a tautology—but neither is it a contradiction. Thus Olivia, if rational, has some positive credence in the proposition that Jekyll is Hyde. All else being equal, then, Olivia's credence in HA might be expected to increase at least slightly on discovering JA, and so we would expect $Cr_2(HD)$ to be greater than 0.2. Of course, it might be that all else isn't equal, but that would need to be spelt out in the example. As it stands, there is no reason to accept that $Cr_2(HD)$ will be exactly 0.2. But Chalmers's more general point—that rationality does not require $Cr_2(HD)$ to be 0.9—does hold.

62   CREDENCE CLAIMS ARE OPAQUE

4.6 Guise Russellianism

So far in this chapter I have given several arguments for the tenet that credence claims are opaque—and the tenet may seem obvious in any case. But as we saw in chapter 2, there is a well-developed and reputable account according to which belief claims are not opaque in quite the way that I have claimed. And so we might wonder: might there be some analogous plausible account of credence claims? In this section I consider and reject this possibility.

I begin by focusing on Salmon's account of belief (Salmon, 1989): other philosophers who hold related views include Scott Soames (Soames, 1987) and David Braun (Braun, 2002). Salmon introduces a three-place relation, BEL. BEL relates a person, a proposition, and a guise. The person is obviously the subject of the BEL attitude, but what, on this view, is a proposition? On Salmon's view, the following two sentences express the same proposition: 'George Orwell is a writer'; 'Eric Blair is a writer'. In general, Salmon's view is that when expressing a proposition, we can substitute a name (like 'George Orwell') for a co-referring name (like 'Eric Blair') without changing the proposition expressed. I note here that Salmon does not hold that names can be safely substituted for co-referring definite descriptions, nor that one co-referring definite description can be substituted for another: the claim is that one name can be safely substituted for another co-referring name without changing the proposition expressed.[10] The final place in the BEL relation is taken by a 'guise'. Exactly what a guise is is a debated and difficult question, and here I do not delve into it closely. As a rough idea, we might think of a guise as something like a Fregean sense: this idea is, of course, itself rather unclear, but an illustration should at least give the general gist. In giving this illustration, I use sentences as a rough stand-in for guises.

To illustrate Salmon's view, let's suppose as usual that Tom works at a café that George Orwell often visits. George Orwell introduced himself to Tom as 'Eric Blair', and Tom is under the impression that this customer Eric Blair is a postman (perhaps because he visits the café early each morning with a large bag of papers) and not a writer. Tom has no idea that the customer Eric Blair who visits his café is the same person as the author

---

[10] We might say that on Salmon's view belief contexts are intensional but not hyperintensional—insofar as that co-referring names (which have the same intensions) can be freely co-substituted, but definite descriptions (which do not have the same intensions) cannot. But Salmon's view is not a version of possible worlds semantics, and to use these terms here could be misleading.

George Orwell. To put things in Salmon's terms, we can describe Tom's epistemic state in terms of BEL relations as follows, where a BEL relation holds between a person, a proposition, and a guise:

BEL (Tom, George Orwell is a writer, 'George Orwell is a writer')

¬BEL (Tom, George Orwell is a writer, 'Eric Blair is a writer')

BEL (Tom, George Orwell is a postman and not a writer, 'Eric Blair is a postman and not a writer')

¬BEL (Tom, George Orwell is a postman and not a writer, 'George Orwell is a postman and not a writer')

In other words, the BEL relation holds between Tom, the proposition that George Orwell is a writer (which is of course the same as the proposition that Eric Blair is a writer), under the 'George Orwell is a writer' guise, but not under the 'Eric Blair is a writer' guise; and the BEL relation holds between Tom, the proposition that George Orwell is a postman and not a writer, under the 'Eric Blair is a postman and not a writer' guise, but not under the 'George Orwell is a postman and not a writer' guise.

Salmon has introduced this three-place BEL relation, but on his view this does not replace the two-place *believes* relation that holds between a person and a proposition, but rather is constitutive of it in the following way: necessarily, the *believes* relation holds between a person and a proposition iff the BEL relation holds between that person and that proposition and *some* guise. Thus, we can see from the BEL relations involving Tom listed above that the following *believes* relations hold:

(4a)   Tom believes that George Orwell is a writer.

(4j)   Tom believes that George Orwell is a postman and not a writer.

Thus the beliefs that Tom has about George Orwell are in conflict. On Salmon's view, this is not a problem. A person's epistemic state is not incoherent just because she believes contradictory propositions: a person's epistemic state is incoherent only if she stands in the BEL relation to contradictory propositions under a guise and the relevant guise-negation, which Tom does not do.[11]

---

[11]   See (Braun, 2016, p. 474) on the idea of a 'guise-negation'.

64   CREDENCE CLAIMS ARE OPAQUE

How does this account fit with our intuitions? Our intuitions seem to pull us both towards and away from this account. On the one hand, it might seem obvious that Tom can believe that Eric Blair is a postman, say, without also believing that George Orwell is a postman. But on Salmon's account this cannot be. For on his view 'Eric Blair is a postman' and 'George Orwell is a postman' express the very same proposition, and the believes relation holds between a person and a proposition, so if Tom believes that Eric Blair is a postman then that simply entails (given that George Orwell is Eric Blair) that Tom believes that George Orwell is a postman. In this way, then, Salmon's account might seem to run counter to our intuitions: on Salmon's view (4k) entails (4l), and yet intuitively in some situations (4k) might be true while (4l) is false:

(4k)   Tom believes that Eric Blair is a postman.

(4l)   Tom believes that George Orwell is a postman.

We have an intuition, then, that (4l) does not follow from (4k), and so our intuitions pull us away from Salmon's account. And yet in other ways our intuitions drive us towards the account. To see this, let's revisit the scenario of Tom, who sees Eric Blair the customer come into the café every morning with his big bag of papers, and has no idea (we would say) that he is George Orwell the author. This scenario was designed to elicit the intuition that (4k) is true and (4l) false, and so if we find that even in this scenario (4l) is true, then that is good evidence that (4k) entails (4l). You might be able to hear (4l) as true straightaway: perhaps you hear it as saying that Tom believes *of George Orwell* that he is a postman, and (you might think) of course this is true in some sense. If you cannot hear this reading at once, then a few interim moves can help.[12] First imagine George Orwell telling a fellow diner about Tom's misapprehension in the following words:

(4m)   Tom believes that I am a postman.

It seems that if (4k) is true, then (4m) (as uttered by George Orwell) is also true. Thus it seems that we can replace the name 'George Orwell' in (4k) by the indexical 'I'—provided that the indexical picks out the right thing—and guarantee that truth is preserved. And if (4m) is true, then the fellow diner

---

[12] These interim moves are inspired by related but different interim moves discussed in (Soames, 1987).

GUISE RUSSELLIANISM 65

can report this fact to a third party by saying (4l). For example, we can imagine George Orwell's fellow diner Jack London telling a mutual friend: 'I sometimes go to the café with George Orwell (not booking under our pen names of course), and Tom the café owner always recognizes me, but he believes that George Orwell is a postman'—and intuitively this statement is true. Thus, given the right conversational context, (4l) has a true reading—even in the scenario that we have constructed specially to elicit the intuition that (4l) is false. Thus it seems that (4k) does entail (4l) after all, and in this way our intuitions push us towards Salmon's account.

You might agree with Salmon that (4k) does entail (4l), but still think that in the Tom scenario there is *something* unsatisfactory about (4l). Salmon's account aims to also explain this dissatisfaction. On Salmon's view, (4l) entails that the BEL relation holds between Tom, the proposition that George Orwell is a postman, and *some* guise or other. All that (4l) entails is (on Salmon's view) true. But it also has an 'implicature', and the implicature is false. 'Implicature' is Paul Grice's term for what is implied but not entailed by a statement (Grice, 1975 (1989)). On Salmon's view, statements attributing a belief in some proposition to a person entail that the BEL relation holds between that person, that proposition, and *some* guise, but they *imply* that the relevant guise is connected with the terms in which the proposition is expressed in the statement. Thus statement (4l) implies that the BEL relation holds between Tom, the proposition that George Orwell is a postman, and the 'George Orwell is a postman' guise, and this implicature is false. Thus Salmon can explain why the statement—though true on his view—strikes us as unsatisfactory: it is because it has a false implicature. Salmon can also explain why our intuitions about the statement vary depending on conversational context, as implicatures are often sensitive to context.

Salmon's account has many virtues, then, and though it has faced criticism, it is a compelling and well-respected theory. I do not discuss the various debates over Salmon's account here. Instead, I consider whether it can plausibly be extended in a particular direction—a direction that Salmon does not himself endorse.[13] The aim is to see whether Salmon's account can be adapted to the credence framework to yield an account which contradicts my tenet. My tenet is that credence claims are opaque, in such a way that we

---

[13] Salmon's own view is that the word 'credence' (as generally used in the literature to mean something like degree of belief) lacks a clear sense: on Salmon's view, belief is typically not a matter of degree (Salmon, 2019, p. 644).

66 CREDENCE CLAIMS ARE OPAQUE

cannot safely substitute co-referring designators—whether names or definite descriptions. On Salmon's account belief claims are not opaque in quite this way: on his view co-referring names can be safely substituted. Could we, then, adapt Salmon's account of belief claims to give an analogous account of credence claims that contradicts my tenet? I turn to this now.

4.7 A guise-based account of credence

A key part of Salmon's account is his three-place BEL relation, which holds between a person, a proposition, and a guise. Could we have a similar relation on the credence framework? This is explored in (Braun, 2016). The natural move here is to propose a four-place CRED relation, which holds between a person, a proposition, a number, and a guise. Thus, for example, we could say that in the Tom scenario the following CRED relations hold:

(4n)   CRED (Tom, George Orwell is a writer, 0.8, 'George Orwell is a writer')

(4o)   CRED (Tom, George Orwell is a writer, 0.2, 'Eric Blair is a writer')

This reflects the fact that—as we might say intuitively—Tom's credence that George Orwell is a writer is high (0.8) under the 'George Orwell is a writer' guise, but low (0.2) under the 'Eric Blair is a writer' guise. On Salmon's view, the belief relation is connected to the BEL relation in the following way: the belief relation holds between a person and a proposition provided that the BEL relation holds between that person and that proposition and *some* guise. We could similarly claim that the credence relation is connected to the CRED relation—but what exactly should the connection be? Perhaps the most obvious analogous move would be to say that an agent stands in the credence relation to a proposition and a number, provided that there is *some* guise such that the agent stands in the CRED relation to the proposition and number and that guise. This move is discussed in (Braun, 2016), and I start by exploring this option.

Under this option, a rational agent can stand in the credence relation to a proposition and lots of different numbers. For the CRED relation might hold between a person, a proposition, a guise, and a number, and also between that same person, the same proposition, and some other guise and other number. To illustrate this, consider again the case of

A GUISE-BASED ACCOUNT OF CREDENCE   67

Tom and claims (4n) and (4o) above, which both hold in the Tom scenario. From claim (4n) claim (4p) follows, and from claim (4o) claim (4q) follows, so claims (4p) and (4q) also both hold in our scenario:

(4p)   Tom has a credence of 0.8 that George Orwell is a writer.

(4q)   Tom has a credence of 0.2 that George Orwell is a writer.

This causes a serious problem for a user of the credence framework. On the credence framework, an agent's credal state can be represented by a credence function which maps propositions onto numbers. To say that it is a function is to say that each proposition is mapped onto at most one number. But if we want to represent Tom's epistemic state using credences, and (4p) and (4q) can both be true, then Tom's epistemic state will not be represented by a credence function, for the same proposition will be mapped to more than one number. This view would require a serious overhaul of the credence framework.

Braun has argued for just such a serious overhaul (Braun, 2016). On his view, we need to distinguish carefully between the objects to which agents bear credence relations, and the objects over which the credence function ranges. The objects to which agents bear credence relations, on Braun's view, are propositions, and the proposition expressed by 'George Orwell is a writer' is the same as the proposition expressed by 'Eric Blair is a writer'. Thus statements attributing credal relations between a person and a proposition are not opaque in the way that I claim: if Tom has a credence of 0.8 that George Orwell is a writer, then it follows that he must also have a credence of 0.8 that Eric Blair is a writer. On the other hand, the objects over which the credence function ranges are *not* propositions on Braun's view: rather they are proposition-guise pairs. Thus Tom's credence function might assign a credence of 0.8 to the proposition-guise pair that consists of the proposition that George Orwell is a writer and the 'George Orwell is a writer' guise, but a credence of 0.2 to the proposition-guise pair that consists of the same proposition and the 'Eric Blair is a writer' guise. This allows Braun to claim that at least some sorts of credence attribution statements are not opaque in the way that I claim, while also claiming that a rational agent's epistemic state can be modelled using a credence function. There are costs to holding this view. One cost is that if the objects over which the credence function ranges are proposition-guise pairs, then it is unclear whether or how they might form an algebra, and so the credence framework may

68 CREDENCE CLAIMS ARE OPAQUE

require a serious rethink.[14] Another cost is that we need to accept a rupture between the objects over which the credence function ranges and the objects to which agents bear credence relations. A rationale for this rupture can be given—for we can see it as analogous to a rupture between the objects of belief and the objects to which an agent might stand in the BEL relation, and this is a rupture endorsed by theorists who accept a guise-based account of belief—but many users of the credence framework will be reluctant to accept this move.

But, in any case, for our purposes the main point to note is that even on Braun's view—which might seem the most promising path to rejecting the tenet—many of the credence attribution statements that we are generally interested in will be opaque. For if we represent an agent's epistemic state by describing her credence function, then in giving some element of the algebra, we will need to give a proposition-guise pair. If, for example, we say that an agent's credence function assigns 0.8 to *George Orwell is a writer*, then the words in italics will need to pick out both the relevant proposition and the relevant guise. We cannot then safely substitute the name 'Eric Blair' for the name 'George Orwell', for to do so would be to point to a different guise, and so a different proposition-guise pair. On this view, then, the tenet is still preserved—insofar as the credence attribution statements that interest us involve representing an agent's epistemic state by describing her credence function.[15,16]

---

[14] In setting out the credence framework in chapter 3, I focused on an approach that involves a state space, an algebra of events over that space, and a function defined on that algebra (Kolmogorov, 1933 (1950)). So understood, the credence framework would require a major overhaul to accommodate Braun's view. As noted however (in footnote 7 of chapter 3), other approaches are possible, and it may be that de Finetti's version of the credence framework (de Finetti, 1931) can accommodate Braun's view more smoothly.

[15] An alternative is to think of any assignment of credences to an agent as an assignment *relative to a guise*. One of the problems with this idea is that we expect to be able to predict, explain, or justify an agent's choice behaviour given her credence and utility functions. Suppose that Tom's credence function relative to one guise suggests (given his utility function) that he will leave the café, but his credence function relative to another guise suggests that he will stay in the café. How can we move from here to a prediction of his behaviour? It's not as though he can leave the café under one guise and stay under another guise! If credences are to play their expected role, then, we should expect an absolute assignment of credences to an agent rather than a range of different assignments under various guises.

[16] A reviewer has suggested the following adaption of Braun's view. To understand it, take the following claim: Tom's credence function assigns 0.8 to *George Orwell is a writer*. The relevant proposition-guise pair (the pair to which Tom's credence function is said to assign a value of 0.8) might naturally be supposed to be fixed by the embedded sentence 'George Orwell is a writer', but on this adapted view, it is instead fixed by the context of utterance. That context itself can be affected by the words of the embedded sentence: the sentence 'George Orwell is a writer' will typically invite a particular 'contextual resolution', which differs from the contextual

4.8 A different guise-based account of credence

Before finishing this chapter, I briefly consider two alternative approaches to extending Salmon's view. On these alternative approaches, we will be able to avoid Braun's rupture between the objects to which agents bear credence relations and the objects over which credence functions range—but as we will see these alternative approaches face other insurmountable problems.

Let's begin by thinking afresh about the connection between CRED relations and credence relations. Recall that on Salmon's view, an agent stands in the belief relation to a proposition provided that there is *some* guise such that the agent stands in the BEL relation to the proposition and that guise. This results in a liberal attribution of beliefs: you count as positively believing a proposition just so long as there is some guise such that you stand in the BEL relation to that proposition and that guise. To take a similarly liberal approach in the case of credences, we might say that the credence that an agent has in a proposition is the maximum number such that there is a CRED relation between that agent, that proposition, and some guise. This is our first alternative guise-based approach. On this view, if (4n) and (4o) hold, and there are no other relevant CRED relations, then Tom's credence that George Orwell is a writer would simply be 0.8, because this is the highest value that stands in a CRED relation with Tom, the proposition in question, and any guise. On this view each person and proposition stands in the credence relation together with at most one number. Thus a person's epistemic state can be represented by a credence function—in that it can be represented by a function that assigns just one number to each proposition in the domain.

The problem with this approach is that we end up having to say that a person—even a rational person—can have a credence function that violates the probability axioms. To see this, consider again the Tom scenario, in which the following CRED relations hold:

resolution typically invited by the sentence 'Eric Blair is a writer'—but were it possible to hold the context fixed, then we could safely substitute 'Eric Blair is a writer' for 'George Orwell is a writer' and expect these sentences to pick out the very same proposition-guise pair. Strictly speaking, on such a view my tenet would not hold for claims about an agent's credence function, for within such claims co-referring names can be safely substituted provided that the context is held fixed. At a practical level, it may make little difference of course: on this view, a change in name will typically result in a change in context, which will then change which proposition-guise pair is picked out, and so it will generally *not* be safe to substitute co-referring names in statements about an agent's credence function—though the explanation for why that is so will need to go via the risk of a change in context. Nevertheless this view may have interesting implications—particularly for the discussion of foundations in chapters 7 and 8.

70 CREDENCE CLAIMS ARE OPAQUE

(4n)   CRED (Tom, George Orwell is a writer, 0.8, 'George Orwell is a writer')

(4o)   CRED (Tom, George Orwell is a writer, 0.2, 'Eric Blair is a writer')

(4r)   CRED (Tom, It's not the case that George Orwell is a writer, 0.2, 'It's not the case that George Orwell is a writer')

(4s)   CRED (Tom, It's not the case that George Orwell is a writer, 0.8, 'It's not the case that Eric Blair is a writer')

Claim (4n) states that the CRED relation holds between Tom, the proposition that George Orwell is a writer, the 'George Orwell is a writer' guise, and the number 0.8. Claim (4r) intuitively follows from this (and can in any case be plausibly assumed in our Tom scenario): if Tom stands in the CRED relation to a proposition, a certain guise, and a certain number x, then presumably he stands in the CRED relation to the negation of that proposition, the relevant guise-negation, and the number $1-x$. Claim (4s) similarly follows from claim (4o). From these claims about the CRED relation—and assuming for simplicity that there are no further relevant claims about the CRED relation—we can infer the following claims about the credence relation:

(4p)   Tom has a credence of 0.8 that George Orwell is a writer.

(4t)   Tom has a credence of 0.8 that it is not the case that George Orwell is a writer.

Claim (4p) holds because 0.8 is the maximum number x such that the CRED relation holds between Tom, the proposition that George Orwell is a writer, some guise, and the number x; and similarly (4t) holds because 0.8 is the maximum number y such that the CRED relation holds between Tom, the proposition that it is not the case that George Orwell is a writer, some guise, and the number y. Thus Tom's credence that George Orwell is a writer, and his credence that it's not the case that George Orwell is a writer, sum to more than 1. We can thus see that on this view Tom's credence function will violate the probability axioms. In general, if we wish to represent the epistemic states of perfectly rational but non-omniscient agents like Tom, we would need to drop the probability axioms. This sort of guise-based account of credence thus turns out to be unfitted for users of the credence framework.

Let us then turn to our second alternative guise-based approach. The approach described in the last paragraph identified an agent's credence in a proposition with the *maximum* number x such that the CRED relation holds between the agent and the proposition and some guise. But now let us consider an alternative whereby we identify an agent's credence in a proposition with the *mean* number x such that the CRED relation holds between the agent and the proposition and some guise. To illustrate, let's take an agent who stands in the CRED relation to some proposition A, guise g, and the number 0.1, and who also stands in the CRED relation to the proposition A, guise g', and the number 0.2: these are the only CRED relations connecting this agent and this proposition. Thus on the view we're considering, the agent's credence in A is 0.15 (the mean of 0.1 and 0.2). On this view, the probability axioms are not violated. I do not attempt to prove this here, but just to illustrate we can consider this agent's credence in not-A. Presumably the agent stands in the CRED relation to proposition not-A, guise g, and the number 0.9, and also stands in the CRED relation to the proposition not-A, guise g', and the number 0.8. Thus the agent's credence in not-A is 0.85 (the mean of 0.9 and 0.8), and thus the agent's credence in not-A (0.85) and the agent's credence in A (0.15) sum to 1—as the probability axioms require.

Nevertheless this approach faces a serious problem when it comes to conditionalization. To extend our example, let us suppose that our agent also stands in the CRED relation to proposition A&B, guise g, and the number 0.1, and also stands in the CRED relation to the proposition A&B, guise g', and the number 0.1: these are the only CRED relations connecting this agent and this proposition. Thus the agent's credence in A&B is 0.1 (the mean of 0.1 and 0.1). Now suppose that the agent learns that A. What is her posterior credence in B? Presumably, it ought to be her prior credence in A&B divided by her prior credence in A, and so it ought to be 0.1 (the agent's prior credence in A&B) divided by 0.15 (the agent's prior credence in A as calculated in the last paragraph), which equals ⅔. But what are the relevant CRED relations after the agent learns that A? Presumably as the agent stood in the CRED relation to proposition A, guise g, and the number 0.1, and in the CRED relation to proposition A&B, guise g, and the number 0.1, on learning that A she ought to stand in the CRED relation to proposition B, guise g, and the number 1 (0.1/0.1). And similarly as the agent stood in the CRED relation to proposition A, guise g', and the number 0.2, and in the CRED relation to proposition A&B, guise g', and the number 0.1, on learning that A she ought to stand in the CRED relation to proposition B, guise g', and the number 0.5 (0.1/0.2). Thus after learning that A, the mean number such that the CRED relation holds between our agent and the proposition B and some guise is the mean of 1 and 0.5, which is

72 CREDENCE CLAIMS ARE OPAQUE

0.75—and on the approach we're exploring, this mean number should equal the agent's credence in B. But as we have seen, the agent's credence in B ought to be ⅔, which is not equal to 0.75. Something has gone awry! The underlying issue is that a conditional probability is a ratio (e.g. $P(B|A) = P(A\&B)/P(A)$), and the mean of a ratio is not necessarily equal to the ratio of the relevant mean values. Thus the approach we are exploring leads to incoherence given the assumption that rational agents update by conditionalization.[17] A user of the credence framework thus cannot adopt this view without dropping conditionalization—a key principle of the credence framework.

4.9 Chapter summary

In this chapter I've argued that credence claims are opaque. I gave three reasons to accept this tenet. Firstly, if we think that there is a relationship (whether logical, causal, or normative) between credences and choice behaviour, then credence claims need to be opaque. Secondly, it is only by accepting that credence claims are opaque that we can avoid the claim that all rational agents are (near) omniscient. And thirdly, a natural application of the credence framework to a case of conditionalization requires that credence claims be opaque.

Besides putting forward these reasons to accept the tenet, I also considered and rejected an alternative option, which is to extend Salmon's guise-based account of belief in the hope of reaching a plausible account of credences on which the tenet is false. Guise-based accounts of credence face an uphill struggle: to make them work, some fundamental principle of the credence framework needs to budge. Braun has offered the most plausible and worked out guise-based account, but on his account the tenet I argue for is largely true: if we represent an agent's epistemic state by describing her credence function, then the credence claims that we use will be opaque in the relevant sense.

Having—I hope—persuaded you of my tenet, I turn now to the part of the book where I trace some of the implications of it. I trace the implications for various principles of rationality in chapter 5, and then some of the practical implications in chapter 6.

---

[17] Thanks to Timothy Williamson (personal correspondence) for discussion on this point.

5
Implications for Rationality

5.1 Introduction

In the last chapter, I argued that credence claims are opaque. In this chapter, I turn to the question: what are the implications for the principles of rationality? Many principles of rationality are put forward, debated, and discussed in the credence framework literature. Probabilism is the principle that a rational agent's epistemic state can be represented by a credence function that obeys the probability axioms; conditionalization is the principle that a rational agent's credence function is updated always and only by conditionalization on evidence gained; the Reflection Principle requires an agent to defer to her future self, and there are also various principles governing peer disagreement; and yet another principle has been proposed—the Principal Principle—which requires agents to defer to the chance function. There are various defences in the literature of these principles, including dutch-book arguments, accuracy arguments, and representation theorems. The central claim of this book—that credence claims are opaque—has implications for all of these principles and the arguments that support them.

To illustrate, consider probabilism, which states that a rational agent's epistemic state can be represented by a credence function that obeys the probability axioms. Our understanding of what a credence function is obviously depends on what sorts of objects are in its domain—i.e. on what sorts of things the objects of credence are.[1] I claim that credence claims are opaque, and so when describing an object of credence it matters which designators we use. Thus Tom can have a credence of 0.8 that George Orwell is a writer, but a credence of 0.2 that Eric Blair is a writer, so in this context 'George Orwell is a writer' and 'Eric Blair is a writer' must pick out different objects of credence. This helps to fix what sorts of things the objects of

---

[1] In chapter 4 we saw that for Braun the expression 'objects of credence' is ambiguous: it may refer to the objects that form the domain of a credence function, or it may refer to the objects to which an individual stands in a credal relation (Braun, 2016). In this chapter, I assume for simplicity that these are one and the same, but for those who endorse Braun's account, my points will be applicable just to 'objects of credence' in the first sense.

The Objects of Credence. Anna Mahtani, Oxford University Press. © Anna Mahtani 2024.
DOI: 10.1093/oso/9780198847892.003.0005

74 IMPLICATIONS FOR RATIONALITY

credence can be: we cannot have an account on which 'George Orwell is a writer' and 'Eric Blair is a writer' pick out the same object. We can summarize this thought by saying that the objects of credence must be *fine-grained*: they must be such that there is a distinction to be drawn between the object picked out by 'George Orwell is a writer' and the object picked out by 'Eric Blair is a writer'.[2]

From the fact that the objects of credence are fine-grained in this sense many further facts follow. For example, on the credence framework each object of credence is a set of states (as described in chapter 3), and if the objects of credence are fine-grained, then this has implications for what these states could be. And there are also knock-on effects for our interpretation of the probability axioms. For example, the probability axioms require that tautologies should be assigned a value of 1—but what counts as a tautology depends on what the objects of credence are, and on the nature of the underlying states. In addition, there are various readings of the dutch-book, accuracy arguments, and representation theorems—and again the fact that the objects of credence are fine-grained will promote some of these readings and rule others out. There are also implications for the principle of conditionalization (and the arguments for it): just as the objects of credence are fine-grained, so evidence must be fine-grained as well.

In this chapter, my aim is simply to demonstrate that the fact that credence claims are opaque has serious repercussions for theorists working in this field. Many of the principles of rationality that Bayesian epistemologists discuss and debate need to be rethought in the light of the fact that credence claims are opaque. I illustrate this by focusing on just two of the principles mentioned in this introduction: the Reflection Principle and the Principal Principle. I turn to the first of these now.

5.2 The Reflection Principle

The Reflection Principle has caused some controversy. There are several different versions of the principle in the literature: for some versions of the principle, counterexamples seem easy to produce, while there are other

---

[2] To be precise, there must be a distinction between the objects picked out by these expressions *when these expressions appear within the context of credence attributions*. I add this qualification because for Frege, for example, the reference of an expression can depend on whether or not it appears in an intensional context: the reference of an expression *when it appears in an intensional context* is its customary sense. See section 2.3.

THE REFLECTION PRINCIPLE   75

versions that seem much more plausible—but for even the most apparently compelling versions of the principle we can still find puzzling counterexamples. I will discuss some of these counterexamples shortly, but I will start by introducing the principle—first in a rough and ready way, and then more precisely.

5.2.1 The Reflection Principle and the Generalized Reflection Principle

Suppose that at t_0 you are about to turn over the top card from a randomly shuffled deck. Your credence at t_0 that (HEART) the top card is a heart is, unsurprisingly, ¼. But what if you could find out what your credence will be in a minute's time at t_1, after the card has been revealed? Let's suppose that this is possible—that somehow at t_0 you learn with certainty what your credence will be at t_1 in HEART (and that you learn nothing else). Surely then it would be rational for you to adopt that credence right away at t_0. For example, if at t_0 you learn that at t_1 your credence in HEART will be 1, then surely you should have a credence of 1 in HEART at t_0; alternatively, if you learn at t_0 that your credence at t_1 in HEART will be 0, then surely you should have a credence of 0 in HEART at t_0. The Reflection Principle endorses this intuition. More generally, the Reflection Principle states that a rational agent defers to her future self.

Reflection Principle: A rational agent defers to her future self.

To explain what is meant by 'defers', we can let Cr_0 stand for an agent's credence at t_0, and Cr_1 stand for that same agent's credence at some particular later time t_1. Then to say that the agent at t_0 defers to herself at t_1 is to say that for any proposition P and value v, such that $Cr_0(Cr_1(P) = v) > 0$:

$$Cr_0(P|Cr_1(P) = v) = v$$

This means that a rational agent's credence in P—conditional on the claim that at the later time her credence in P will be v—is v. It follows from this that if (as in our HEART example above) a rational agent at t_0 is *certain* that her credence in P at t_1 will be v, then her credence in P at t_0 is v. There are also consequences in cases where a rational agent is not certain what her future credence in P will be: in such cases her credence in P at t_0 ought to

76 IMPLICATIONS FOR RATIONALITY

equal her expectation of her future credence in P. An agent's expectation of some variable is a sort of weighted average: in this case, to calculate the agent's expectation of her future credence in P, we would take each value that the agent's future credence might take, multiply it by the agent's credence that her future credence will indeed take that value—and then we sum the results.

On the face of it, the principle has a certain intuitive pull, and it is also backed up by a dutch-book argument—though there are various reasons to question the force of this argument (Briggs, 2009b; Mahtani, 2012). But the principle is certainly not entirely satisfactory, for it faces various counterexamples. For one such counterexample, take a case where you suspect you might forget something in the future: perhaps you're confident that you ate spaghetti for dinner today, but you suspect that in a year's time you'll have forgotten this, in which case intuitively you should not defer to your future self over what you had for dinner today (Talbott, 1991). For another counterexample, take a case where you suspect that you might become deluded or irrational in the future: again, intuitively you should not then defer to your future self (Christensen, 1991). One reply that could be made in the face of these counterexamples is that no completely rational agent would forget something or become deluded or irrational in the future. And perhaps this is right, given the technical, idealized sense of 'rational' that Bayesians often work with. To deal with our counterexamples, indeed, the Bayesian would need to go even further, and claim that no rational agent would even *suspect* that she might become forgetful, deluded, or irrational—and, indeed, that all rational agents are completely certain that this will never happen—but perhaps even this can be endorsed given a stringent enough sense of 'rational' (Sobel, 1987).[3] Still, it would be useful to have a restricted version of the principle that can apply even to agents who are not rational in this extreme idealized sense.

Here is one plausible restriction. In some cases, an agent at t_0 'respects' her future t_1 self in the following sense: at t_0 she is quite certain that the only way in which she will change between t_0 and t_1 is that she may gain some evidence, and that if she does so, then she will simply conditionalize on it. We can then give a much more compelling version of the principle:

---

[3] See (Green and Hitchcock, 1994) for an argument that the constraints that the Reflection Principle places on an agent are less stringent than generally supposed.

THE REFLECTION PRINCIPLE 77

Restricted Reflection Principle: If a rational agent respects her future self, then she defers to that future self.[4]

And this suggests a way to broaden the principle out, for it seems that it is the respecting relation that is doing the work here, and it shouldn't matter whether the respected agent is one's future self or somebody else. Let us say that an agent (at a time) respects an(other) agent (at a time) iff the first agent is certain that the second agent's credence function is identical to her own, except that the second agent may have acquired some extra (true) evidence, in which case (the first agent is sure) the second agent will have simply conditionalized on that extra evidence. Thus we can give a generalized version of the Reflection Principle as follows:

Generalized Reflection Principle: If a rational agent-at-a-time A respects an agent at-a-time A*, then A defers to A*.

This generalized principle seems like an improvement on the original principle. However, strangely, it still faces counterexamples. Here is a case to illustrate this (Mahtani, 2016).

5.2.2 The Mug

This situation involves a dealer, three players (Alice, Bob, and Carol), and a pack of cards. Let us assume that the three players begin with all the same relevant evidence and the same credences in all relevant propositions, that they all know this, and further that they all know that they are all rational and will change if at all only by conditionalizing on evidence gained. The dealer randomly and secretly selects one player to be 'the mug' and one to be 'the lucky player', and then the remaining player is simply 'the other player'. The dealer deliberately deals a joker to the mug, an ace to the lucky player, and deals either a joker or an ace—chosen at random from a deck consisting of just aces and jokers in equal proportion—to the other player. All the players understand this set-up, but none knows how the roles (the mug, the lucky player, the other player) have been assigned. Once the cards have been dealt (but not examined) the time is t_0; then at t_1 each player looks

---

[4] This restricted reflection principle can be shown to follow from the probability axioms given some 'idealizing assumptions' (Briggs, 2009b, pp. 69–70).

78 IMPLICATIONS FOR RATIONALITY

privately at the card that (s)he has been dealt. Throughout players consider the following proposition:

TWO ACES: Two of the cards dealt out are aces.

Let us start by considering, say, Alice's epistemic state at t_0. Her credence in TWO ACES at this point should be ½. Alice considers her future epistemic state: what will her credence in TWO ACES be at t_1? Alice at t_0 respects her t_1 self: she knows that by t_1 she will have gained some evidence by looking at her card, and that she will have simply conditionalized on this evidence. Thus the Reflection Principle—including the generalized version discussed above—mandates that she should defer to her future self. At t_0 Alice has a credence of ½ that her t_1 self will see that she has an ace (in which case her new credence in TWO ACES will be ⅔) and a credence of ½ that her t_1 self will see that she has a joker (in which case her new credence in TWO ACES will be ⅓)[5] and so her expectation of her future credence is $(½ × ⅔) + (½ × ⅓) = ½$. So far so good: her current credence in TWO ACES matches the expectation of her future credence in TWO ACES, as the Reflection Principle demands. But now suppose that Alice at t_0 considers the epistemic state of the mug at t_1. Right now, at t_0, Alice knows that she and the mug have all the same relevant evidence, and are effectively in the same epistemic state. Alice also knows that at t_1, after turning over his or her card, the mug will have gained some evidence and will simply conditionalize on it. Thus Alice at t_0 respects the mug at t_1 and so our generalized version of the Reflection Principle requires Alice at t_0 to defer to the mug at t_1. It follows that Alice's credence at t_0 in TWO ACES ought to equal her expectation of the mug's credence at t_1 in TWO ACES. But Alice already knows what credence the mug will have at t_1 in TWO ACES: Alice knows that the mug will see that (s)he has been dealt a joker, and so will have a credence in TWO ACES of ⅓. It seems, then, that the generalized version of the Reflection Principle requires Alice at t_0 to have a credence of ⅓ in TWO ACES. This doesn't seem right! And, indeed, it can't be right, because by similar reasoning our Generalized Reflection Principle would also require Alice at t_0 to defer to the Lucky Player at t_1, and so have a credence of ⅔ in TWO ACES, and she can't have both a credence of ⅓ and a credence of ⅔ in TWO ACES at the same time. What has gone wrong?

---

[5] The relevant calculations are set out in the next paragraph.

THE REFLECTION PRINCIPLE   79

To run through the calculations here, we can set out Alice's credences using the following table containing six possible states. At t_0, Alice has a credence of 1/6 in each:

I am the mug, and the 'other player' is dealt a joker	I am the mug, and the 'other player' is dealt an ace
I am the lucky player, and the 'other player' is dealt a joker	*I am the lucky player, and the 'other player' is dealt an ace*
I am the 'other player', and the 'other player' is dealt a joker	*I am the 'other player', and the 'other player' is dealt an ace*

TWO ACES obtains in every cell of the column on the right, and so at t_0, Alice has a credence of ½ in TWO ACES. Alice is dealt an ace iff she is at one of the states in italics, and so at t_0 Alice has a credence of ½ that she is dealt an ace—and similarly a credence of ½ that she is dealt a joker. At t_1, Alice finds out whether she has been dealt an ace or a joker. If she finds out that she has been dealt an ace, then she can eliminate all states other than those in italics, and she will then have a credence of 2/3 in TWO ACES. For analogous reasons, if she finds she has been dealt a joker, then she will have a credence of 1/3 in TWO ACES. The mug—whoever that is—will reason similarly, and so on seeing a joker at t_1 will have a credence of 1/3 in TWO ACES; and the lucky player—whoever that is—will have a credence at t_1 of 2/3 in TWO ACES. Alice knows all of this at t_0, and thus cannot defer to both the mug at t_1 and the lucky player at t_1.

This case is closely related to a cluster of other cases discussed in the literature, including the Sleeping Beauty problem (Elga, 2000), the prisoner cases (Arntzenius, 2003), the Cable Guy paradox (Hájek, 2005a), and the puzzle of the hats (Bovens and Rabinowicz, 2010). Many different approaches have been put forward in response to these cases, and I discuss some of these below. But I believe that the root of the problem here is a failure to recognize that credence claims are opaque. I turn to explain now how this point relates to the puzzle at hand, and to propose a new version of the Reflection Principle that accommodates this point.

5.2.3 The Improved Generalized Reflection Principle

The key point to recognize is that just as credence claims are opaque, so are deference claims. It can be true both that you defer to some individual under

80 IMPLICATIONS FOR RATIONALITY

some way of designating her, and that you don't defer to the very same individual under some other way of designating her. We can see an intuitive example of this by thinking about our case of Tom the café worker. He may consider George Orwell to be an expert on literature, and defer to him on all such matters, while also considering Eric Blair to be very ignorant, and not deferring to him at all. To illustrate: if Tom were to learn that George Orwell has a high credence that metaphors are better than similes, then Tom would immediately have such a high credence himself; whereas if Tom were to learn that Eric Blair has a high credence that metaphors are better than similes, then Tom would barely adjust his credence at all. Just as credence claims are opaque, so deference claims are opaque too. We can say, then, that the objects of deference are fine-grained: deference is not a relation between two individuals, but rather between an individual (Tom in our case) and an individual *under a designator* (George Orwell so designated, in our example, but not Eric Blair so designated). This point applies in cases of deference to oneself. I might defer to my future self so designated, but there may be a designator that applies to my future self (the person at particular co-ordinates at a particular future time) under which I would not recognize myself, and so I may have no reason at all to defer to the individual so designated.

Once this is recognized, then we can see that the Reflection Principle— including our generalized version—needs to be rethought. We cannot require simply that a rational agent defers to any *individual* who meets certain criteria—for deference is not a relation between two individuals, but rather a relation between an individual and a fine-grained object—an individual *under a designator*. How then should the Reflection Principle be understood? As it stands the definition is incomplete: we are told to defer to an individual who meets certain criteria—but under what designator(s)? Requiring deference to an individual who meets the criteria under *all* that individual's designators is far too demanding: clearly I do not need to defer to my future self under a designator that I do not even recognize as applying to myself.

To handle this problem, a first point to note is that claims about 'respecting' (as I have defined the term) are also opaque, and so the objects of respect are similarly fine-grained: an agent might respect an individual under one designator without respecting that same individual under a different designator. This follows directly from the fact that the objects of credence are fine-grained. Tom may be certain that George Orwell knows all that he (Tom) knows and more, while doubting whether Eric Blair has a similar

THE REFLECTION PRINCIPLE 81

body of knowledge. Thus we can clarify the Generalized Reflection Principle by adding the bracketed clauses in the principle below:

Clarified Generalized Reflection Principle: If a rational agent-at-a-time A respects an agent at-a-time A* (so designated), then A defers to A* (so designated).

Thus if Tom respects George Orwell (so designated), then if rational he defers to George Orwell (so designated)—but it doesn't follow that he defers to Eric Blair (so designated). This gives us a useful clarification of the Generalized Reflection Principle. But further adjustments are needed: to get a plausible version of this principle we need to carve out a category of appropriate designators. I explain why this adjustment is needed using once again the example of the mug.

In our example of the mug, we have three players, but each player can be designated in more than one way. For we can designate the three players using the following set of designators {'Alice', 'Bob', 'Carol'}, or equally we can designate the three players—not necessarily in the same order—as {'the mug', 'the lucky player', 'the other player'}. The players can also all designate each other in these ways.[6] Alice at t_0 respects all the players under all of these designators at t_1. For example, Alice at t_0 respects Bob at t_1, for she is certain that Bob at t_1 will have (effectively) the same credence function as she has at t_0, except that he will have gained some further information and conditionalized on it. And similarly, Alice at t_0 is certain that the mug at t_1 will have (effectively) the same credence function as she has at t_0, except that (s)he will have gained some further information and conditionalized on it. Thus given the Generalized Reflection Principle as we have formulated it so far, Alice at t_0 will, if rational, defer to Alice, Bob, Carol, the mug, the lucky

---

[6] It is certainly the case that the players can all competently use all of these designators. For example, they can say (or think), 'Alice may have an ace', and 'the mug definitely has a joker'. But it is tempting to say that there is something defective about the players' use of 'the mug', 'the lucky player', and 'the other player', because when they use these expressions they do not know to whom they are referring. Let us say that when you do not know to whom a designator refers, then for you that is not an 'identified designator'. The implied distinction between identified and non-identified designators turns out to be elusive: what is required, exactly, for a designator to be identified? If the requirement is, for example, that you need to know what the person looks like, or be able to point at them, then there will be ways of setting up our scenario so that you do have the relevant information about 'the mug', 'the lucky player', and 'the other player', and yet the same issues remain. Once the idea of identified designators is made explicit, it turns out to be unhelpful. This idea resurfaces in chapter 6.

82 IMPLICATIONS FOR RATIONALITY

player, and the other player all at t_1, and all so designated. This is impossible, as we have seen: Alice at t_0 cannot consistently defer to both the mug at t_1 and the lucky player at t_1. We need a way to eliminate at least some of these designators as inappropriate.

One promising approach is to distinguish between designators that are 'self-identified' and those that are not, where we define a designator as self-identified iff the individual so designated recognizes (with certainty) that she falls under that designator.[7] How can we use this distinction to carve out a set of designators under which it would be appropriate for one agent at a time (A) to defer to some other agent at a time (A*)? We might try requiring that 'A*' is a self-identified designator—that is, that 'A*' is actually a self-identified designator, whether the agent A realizes it or not—but that is not a strong enough restriction in itself.[8] We might try requiring additionally that for a designator 'A*' to be such that it is appropriate for A to defer to A* so designated, A must be *certain* that 'A*' is a self-identified designator. But this requirement is too restrictive and will exclude some designators unnecessarily. A better requirement is that A must be certain that either A* is a self-identified designator, or if A* were to become a self-identified designator (i.e. if A* were to learn that she is A*) then A*'s credence in the proposition under consideration would not change. Note that on this definition, whether a designator A* is appropriate is relative—relative both to the agent doing the deferring (A) and to the proposition under consideration (P). With all of this in mind, we can give the following improved version of the Generalized Reflection Principle:

Improved Generalized Reflection Principle: If a rational agent-at-a-time A respects an agent at-a-time A* (so designated), and if A* is an appropriate designator, then A defers over proposition P to A* (so designated).

---

[7] While the distinction between identified and non-identified designators is elusive (see footnote 6), it is (relatively) clear what is required for a designator to be self-identified: the referent must be certain that the designator refers to herself.

[8] To see this, we can imagine a variant on the mug case where it is part of the set-up (and so known to all players) that the dealer will secretly toss a 20-sided die, and iff it lands with the 1 face up then the dealer will write 'you are the mug' on the face of the card that she deals to the mug—and so the mug will learn at t_1 that (s)he is the mug. If the die does happen to land with the 1 face up, then the 'the mug' will indeed be a self-identified designator at t_1, and the mug's credence in TWO ACES at t_1 will be ½. But even if in this way 'the mug at t_1' is a self-identified designator, we still should not require the players at t_0 to defer to the mug at t_1, because the players at t_0 will not be certain that 'the mug at t_1' is a self-identified designator, and their expectation of the mug's credence at t_1 in TWO ACES will not equal ½.

THE REFLECTION PRINCIPLE 83

To recap: the designator A* is appropriate iff A is certain that either a) A* is self-identified, or b) if A* were to become self-identified (i.e. if A* were to learn just that she is A*) then A*s credence in the proposition under consideration (P) would be unchanged.

This principle gives us the right result in the case of the mug. For example, Alice at t_0 ought to defer to Bob at t_1 over TWO ACES, for Alice at t_0 respects Bob at t_1, and 'Bob' is an appropriate designator: we can set the example up so that Bob knows his own name (and is known to do so)—in which case 'Bob' is known to be a self-identified designator; or even if we set things up so that Bob does not know his own name, then 'Bob' is an appropriate designator, because if Bob were to learn his own name, then this would not change his credence in TWO ACES, and Alice knows this. Thus Alice at t_0 ought to defer to Bob at t_1, and this is unproblematic: for Alice's credence in TWO ACES at t_0 (½) will match her expectation of Bob's credence in TWO ACES at $t_1((½ \times ⅔) + (½ \times ⅓) = ½)$. A problem would arise if our principle required Alice at t_0 to defer to the mug or the lucky player at t_1. Fortunately, it does not require this, for these are not appropriate designators. Alice cannot be sure that 'the mug' is a self-identified designator; furthermore, she cannot be sure that if the mug were to learn that (s)he is the mug, then (s)he would not change his or her credence in TWO ACES. Indeed, as we set up the original example, Alice can be sure that the mug *would* change his or her credence in TWO ACES from ⅓ to ½ on learning that (s)he is the mug. Thus 'the mug' is not an appropriate designator, and for similar reasons 'the lucky player' is not either, and so our principle does not require Alice to defer to either of them.

An interesting question arises here: *why* is there this connection between deference and self-identified designators? What is the underlying mechanism here? I've puzzled over this question a good deal. To take the example above, I was at first tempted to say that Alice knows something that the mug does not—for the mug does not know that (s)he is the mug—and so Alice should not respect the mug (so designated) after all. But that doesn't seem right, because what does Alice know exactly? Alice knows that the mug is the mug, but the mug has this tautological piece of information too. The mug does not know that (s)he him- or herself is the mug—but then of course Alice doesn't have that piece of indexical information either. It really does seem to be the case that the mug knows all that Alice knows (and more) and so is worthy of respect. Nevertheless, Alice would know that it is the mug to whom she is deferring, and she would also know—or suspect—that the mug would not defer to him- or herself so designated: on learning that (s)he is the

84 IMPLICATIONS FOR RATIONALITY

mug, (s)he would adjust her credence. Why, then, should Alice defer to the mug's credences without any such adjustment? This I think is the underlying thought that motivates the Improved Generalized Reflection Principle.

Thus we have arrived at a version of the Reflection Principle which avoids our counterexample of the mug. This version can also handle the many other similar puzzling cases discussed in the literature. For many of these cases, we need to look carefully at the way that the time (rather than the person) is designated. Here is one famous example to illustrate this.

5.2.4 Sleeping Beauty

It is Sunday night, and Sleeping Beauty—who let us assume is fully rational—is about to take part in an experiment. She will be put to sleep, and then woken up again on Monday. She will then be put back to sleep, and in the afternoon a coin will be tossed. If it comes up heads, then she will be left to sleep until she wakes up naturally on Wednesday when she will immediately be told that the experiment is over. If it comes up tails, then her memory of the Monday awakening will be wiped and she will be woken up again—as if for the first time—on Tuesday morning, before being put back to sleep until she wakes up naturally on Wednesday and is immediately told that the experiment is over. Sleeping Beauty considers the proposition HEADS that the coin tossed on Monday night will land heads up. On Sunday night it seems obvious that her credence in HEADS should be ½. What should it be on Monday morning? Many have argued that it should be less than ½ (specifically that it should be ⅓). Compelling arguments for this can be found in, for example, (Elga, 2000) and (Dorr, 2003), but a quick intuitive way to see why this view might be appealing is to recognize that on waking on Monday morning, were Sleeping Beauty to be informed that it was Monday, then her credence in HEADS would presumably be ½ (because she would then know that the coin had yet to be tossed), and so *before* being informed that it is Monday her credence in HEADS must presumably be less than ½, because the only other possibility compatible with her evidence at that point is that it is Tuesday, in which case the coin must have landed tails.

Thus it seems that on Sunday night Sleeping Beauty's credence in HEADS is ½, and yet she can calculate on Sunday night that on Monday morning her credence in HEADS will be less than ½. This violates the Reflection Principle. It also violates the Generalized Reflection Principle, for Sleeping Beauty on Sunday night respects her Monday morning self: she knows that

THE REFLECTION PRINCIPLE   85

she will not forget anything between Sunday and Monday, and that she will remain rational, so if her credence changes presumably this will be because she has acquired some further evidence and conditionalized on this.[9] Thankfully there is no violation of the Improved Generalized Reflection Principle, and this is because Sleeping Beauty *on Monday morning* is not an appropriate designator. To see why, consider that on Monday morning Sleeping Beauty will not know with certainty that it is Monday morning, and so 'Sleeping Beauty on Monday morning' is not a self-identified designator; and furthermore, if on Monday morning Sleeping Beauty were to learn that it is Monday morning, then her credence in HEADS would change (from less than ½ to ½). Sleeping Beauty on Sunday night can calculate all of this, and so is not required to defer to Sleeping Beauty on Monday morning.

This response is closely related to that of (Schervish, Seidenfeld, and Kadane, 2004). The Reflection Principle states that an agent, if rational, will defer to herself at any future time: to use the terminology of Schervish et al., the relation is between an agent *now* and the same agent *later* (with *now* and *later* designating times). Schervish et al. claim that the Reflection Principle holds—provided that certain requirements are met. The one that interests us here is this: at *now*, the agent must be certain that either *later* is a stopping time (i.e. that at any given time she will know whether or not it is currently *later*), or that learning at *later* that it is *later* will not affect her assessment of the relevant claim. My concept of an appropriate designator differs from the concept of a stopping time in two ways. Firstly, it is a generalization of that idea. As well as 'stopping times' we can also have what we might call 'stopping people', where a person is a stopping person iff she recognizes herself as that person (so designated) with certainty. Thus in our example of the mug, Alice is a stopping person (assuming that she knows her own name), but the mug is not a stopping person. Secondly, and more importantly for the purposes of this book, my concept of an 'appropriate designator' clarifies that the distinction applies not to times (or people)—but to ways of designating those times (or people). To put this into the language

---

[9] It is, however, difficult to state exactly what evidence Sleeping Beauty could have gained from waking up on Monday morning. Furthermore, it may be tempting to say that though Sleeping Beauty can be confident that her memory will not be interfered with by Monday morning (any such interference will take place later that day), nevertheless she loses some information between Sunday night and Monday morning, because by Monday morning she no longer knows what day it is. Both of these points connect with a general issue here over how rational agents update indexical claims—such as claims about what day and time it is *now*. I do not delve into this interesting issue in this chapter, but see (Titelbaum, 2016) for an overview.

86 IMPLICATIONS FOR RATIONALITY

of stopping times: it should be recognized that a stopping time is not really a *time*, but a *time under a designator*. For times—just like people—can be designated in different ways. It may be that the time at which the final bus leaves the station is 4 o'clock exactly, and so '4 o'clock' and 'the time when the final bus leaves the station' may designate exactly the same time. Yet it may be that 4 o'clock is a stopping time (relative to an agent—who let's assume is looking closely at a reliable clock), while the time when the final bus leaves the station is not a stopping time (relative to the very same agent, who has no reliable bus timetable, but is—as stated—looking closely at the reliable clock). For these reasons, I use the term 'appropriate designator' in preference to the term 'stopping time' to state the scope of the Reflection Principle.

This concludes the main discussion of the Reflection Principle. To summarize, I have argued that just as credence claims are opaque, so (for closely related reasons) deference claims are opaque. In light of this, the Reflection Principle needed to be rethought: deference is not a relation between two individuals (at times), but rather between an individual (at a time) and an individual (at a time) under a designator. To accommodate this thought, I offered an improved version of the Generalized Reflection Principle, which limited the application of the principle to those designators which are appropriate (in a sense explained). This improved version can handle a range of tricky counterexamples in the literature.

5.2.5 Other principles of deference and disagreement

I end this section by gesturing towards some further principles which need rethinking along similar lines. We have been focusing on principles of deference which govern cases where one agent treats another as an expert, but there are also principles of disagreement which govern cases where one agent treats another as a peer. Much of the debate in the peer-disagreement literature revolves around the question of how agents ought to respond on finding out the credences of their peers. Some hold that were S_1 to learn that her peer S_2 had a credence in some claim P which differed from S_1's own credence, then S_1 ought to 'split the difference'—that is, assign a credence to P that is the average of S_2's and her (previous) own credences (Feldman, 2006; Elga, 2000; Christensen, 1991). Some maintain that S_1 ought to hold steadfastly to her own credence (van Inwagen, 1996; Kelly, 2005). And other more nuanced approaches are also discussed in the literature. For the purposes of this chapter, the point to highlight is that the relation of 'perceived peer-hood' is not a relation between individuals, but rather between individuals under designators. Thus S_1 might consider S_2 a peer under one designator but not under another. And—just as for deference principles—we must impose a restriction to admissible designators if the principles of peer disagreement are to be plausible. Just like principles of deference, then, principles of disagreement similarly need to be reconsidered in the light of the fact that credence claims are opaque.

In the next section, I turn to a different principle of rationality: the Principal Principle. Like the Reflection Principle, this is a principle of deference, and it is also affected by the fact that credence claims are opaque, but the issues here are rather different.

5.3 The Principal Principle

The Principal Principle draws a connection between credences—which are the focus of this book—and chance. Chance plays an enormously important role in many different domains, including science—both natural and social—and it is a very important concept in ethics: to give just one example, some theorists claim that where there is limited access to a resource, and where certain other considerations (such as much greater need) do not apply, then the distribution of that resource should be decided by chance (Diamond, 1967; Broome, 1984). This chapter focuses on principles of rationality rather than on practical concerns, but much of the discussion here can be applied and extended to these wider ethical questions.

I begin by outlining the Principal Principle. The original Reflection Principle, discussed above, states that a rational agent defers to her future self. The Principal Principle, in contrast, states that a rational agent defers to the chance function. As above, we can note here that we have an opaque context: the Principal Principle requires (or should require) you to defer to the chance function *so designated*, rather than to that function no matter how it is designated—but no confusion seems to have arisen over this point. Nevertheless, the fact that credence claims are opaque does create a problem for the Principal Principle. For if credence claims are opaque, then are chance claims opaque too? If credence and chance claims are different in this respect, then this leads to a problem for the Principal Principle, as we shall see.

88 IMPLICATIONS FOR RATIONALITY

Throughout this book I have generally been using the term 'opaque', partly because it captures the phenomenon vividly, and partly because it is not bound up with any particular account of propositions. For the purposes of this section on the Principal Principle, however, it makes sense to draw from a group of terms used in possible worlds semantics: 'extensional', 'intensional', and 'hyperintensional'. This is because these terms help us to draw distinctions that would be difficult to express just with the idea of opacity. In the next section I recall (from chapter 2) how these terms work, and argue that chance claims are not extensional.

5.3.1 Chance claims are not extensional

Designators (along with other expressions) have both extensions and intensions. The extension of a designator is the object that it refers to—so two co-referring designators have the same extension. The intension of a designator is a function from possible worlds to objects—so two designators have the same intension iff they pick out the same object at each possible world. Names are rigid designators, so two co-referring names have the same intension: if two names refer to the same object in the actual world, then the object that they refer to at each possible world must also be the same. Definite descriptions, on the other hand, have a non-rigid reading, so two co-referring definite descriptions (or a name and co-referring definite description) are not guaranteed to have the same intension.

A context can be extensional, intensional, or hyperintensional. In an extensional context, we can safely substitute any designator for any other designator with the same extension without any risk of changing the truth-value of the statement. A sentence like 'Mount Everest is over 8000 metres tall' is extensional in this way: we can harmlessly replace 'Mount Everest' with another name for the same object (e.g. 'Chomolungma'), or with a definite description (e.g. 'the highest mountain above sea level on earth') and the truth-value remains the same. At the other extreme, we have hyperintensional contexts—which I have generally been calling 'opaque' contexts. In a hyperintensional (or opaque) context, substituting one designator for another can change the truth-value of the statement—even if the two designators have both the same extension and the same intension. I have argued that credence claims create such contexts: it may be true that Tom has a credence of 0.8 that George Orwell is a writer, but not true that

THE PRINCIPAL PRINCIPLE   89

Tom has a credence of 0.8 that Eric Blair is a writer. Here by substituting 'Eric Blair' for 'George Orwell' in the credence claim, we have changed the truth-value of the claim, even though 'Eric Blair' and 'George Orwell' have the same extension and the same intension.

Between these two extremes we have contexts that are intensional without being hyperintensional. In these contexts, designators can be safely substituted provided that they have the same intension. Thus co-referring names can be safely substituted, but co-referring definite descriptions cannot be, nor can a name be safely substituted for a co-referring definite description or vice versa. Certain sorts of modal context are generally taken to be like this—that is, intensional without being hyperintensional. For example, the operator 'it is necessary that...' creates a modal context in the following (true) sentence: it is necessary that Mount Everest is Mount Everest. In this context, we can safely substitute the co-referring name 'Chomolungma' for 'Mount Everest', to give us the following sentence: it is necessary that Mount Everest is Chomolungma. Given that 'Mount Everest' and 'Chomolungma' have the same intension—that is, they refer to the same object at every possible world—it is indeed necessary that Mount Everest is Chomolungma, so the truth-value of the sentence is retained. But if we instead substitute for 'Mount Everest' the co-referring definite description 'the highest mountain on Earth', then we change the truth-value of the sentence, for it is not necessary that Mount Everest is the highest mountain on earth: there are possible worlds where a different mountain is higher. Thus this sort of modal context is intensional but not hyperintensional: we can safely substitute co-referring proper names, but not co-referring definite descriptions.

With these distinctions in mind, we can consider what sort of context the chance operator creates. I begin with an example designed to show that names cannot be safely substituted for co-referring definite descriptions— and so that at any rate chance claims are not extensional. Suppose that twenty-six people (Alice to Zebedee) have entered a lottery. The winner of the lottery will be selected tomorrow by a genuinely indeterministic process, so today the chance that each person will win the lottery is just $\frac{1}{26}$. Thus the chance of each of (5a)–(5z) below is $\frac{1}{26}$:

(5a)   Alice will win the lottery.

(5b)   Bob will win the lottery.

...

(5z)   Zebedee will win the lottery.

90 IMPLICATIONS FOR RATIONALITY

Now consider that *somebody* is going to win the lottery: we don't know who it is, of course, and it is currently indeterminate who it will be—but somebody is going to win.[10] What, then, is the chance that the person who will win the lottery will win the lottery?

(5*)   The person who will win the lottery will win the lottery.

There is a reading of (5*) such that the chance of (5*) is 1.[11] Now suppose that it is in fact Alice who will go on to win the lottery: we could make the same point whoever the winner turns out to be. If it is in fact Alice who will go on to win the lottery, then 'Alice' and 'the person who will win the lottery' designate the very same person—that is, they have the same extension. In that case, (5*) differs from (5a) only in having a different designator for this person—and yet the chances of (5*) and (5a) are different. This shows that as far as chance claims are concerned, the truth-value can change when we substitute one designator for another with the same extension, and so that chance claims are not extensional.

Tim Crane and Hugh Mellor make this point, arguing as follows (with '$p(X) = n$' meaning that the chance of X is n):

> ... '$p(\dots) = n$' is not extensional: if it were, 'a is the F' and the necessary truth '$p(a$ is $a) = 1$' would entail '$p(a$ is the $F) = 1$', which it clearly does not, on any view of probability (take for example 'F' = 'next Prime Minister').
>
> (Crane and Mellor, 1990, p. 195)

We can fill out this example by letting 'the F' stand for 'the next prime minister of the UK' as suggested, and letting 'a' stand for Keir Starmer. We can also suppose that (though no one presently knows this, and indeed it is not determined by the current state of the world) Keir Starmer will in fact be the next prime minister of the UK. Thus 'Keir Starmer' and 'the next prime

---

[10] Philosophers who might dispute this claim include certain sorts of presentists and growing block theorists. In this chapter I do not attempt to argue against these views, but just proceed on the (not universally accepted) assumption that there are truths about the future—even when the future is indeterminate.

[11] Arguably there is also a reading of (5*) on which its chance is $1/26$: on this reading, we take 'the person who will win the lottery' to mean the same as 'the person who will *actually* win the lottery'. The point in the main text goes through as long as there is some reading of (5*) on which its chance is 1: on such a reading, we cannot safely substitute a name for a co-referring (non-rigidified) definite description within the context of a chance claim.

THE PRINCIPAL PRINCIPLE 91

minister of the UK' refer to the same object. Crane and Mellor claim—surely plausibly—that the chance that Keir Starmer is Keir Starmer is 1. If chance claims were extensional—that is, if co-referring names and definite descriptions could be freely substituted within the context of a chance claim—then it would follow that the chance that Keir Starmer is the next prime minister of the UK is 1. Crane and Mellor claim that this is clearly false, and so draw the conclusion that the chance context is not extensional. This claim has wide-reaching implications.

One important implication is for the relation of causation. Often causation is assumed to be a relation between particulars: we think, for example, of one event causing another. Donald Davidson, most prominently, sees causation as a relation between two events, and Davidson is explicit in his claim that the relation holds between these events no matter how they are described (Davidson, 1967). But many see causation as closely related to chance: Mellor, for example, sees the claim that a causes b as closely related to the claim that a raises the chance that b (Mellor, 1994). Given this close relation, if chance claims are not extensional, then causation claims must be similarly not extensional. For suppose that a single event can be designated both as 'a' and as 'a^*', and some other event can be designated both as 'b' and as 'b^*'. As chance claims are not extensional, it might be true that a raises the chance that b, even while it is false that a^* raises the chance that b^*, and so given the tight connection between chance and credence it seems to follow that a might cause b even while a^* does not cause b^*: that is, it seems to follow that causation claims are similarly not extensional, and so that Davidson's position is incorrect. Crane and Mellor draw out important and wide-ranging implications of this for the philosophy of mind, and for questions over the appropriate logic for physics. In summary—the claim that chance claims are not extensional has serious ramifications for a range of fields.

Amongst all of this upheaval, it might be supposed that at least the Principal Principle is safe.[12] After all, if neither credence claims nor chance claims are extensional, then plausibly they are similarly opaque and there should be no difficulty drawing a connection between them. That doesn't follow, unfortunately. For though neither credence claims nor chance claims are extensional, there may still be a distinction to be drawn. Credence claims are hyperintensional, but on a popular way of thinking about chances, chance claims are intensional but not hyperintensional. I turn to this way of thinking about chances in the next section.

---

[12] This seems to have been Mellor's view (personal correspondence).

92 IMPLICATIONS FOR RATIONALITY

5.3.2 The chance framework

We begin by considering the set of all metaphysically possible worlds. Each metaphysically possible world has an entire history—past, present, and future. Some of these metaphysically possible worlds have histories that are identical with that of the actual world up till the present moment. We can say that these are the worlds that are 'still in play'. Whatever holds at all of these worlds, holds *determinately*.[13] Thus, for example, at the actual world, the French Revolution occurred in the past, and so this occurred at all worlds that are still in play at the current time. We can say, then, that the French Revolution occurred determinately. In contrast, if we suppose that our lottery winner will be selected at a future time using a genuinely indeterministic process, then it is currently indeterminate who the winner will be: at some worlds in play the winner is Alice; at others it is Bob, and so on. Whatever holds determinately (i.e. at every world in play), holds with chance 1; whatever determinately doesn't hold (i.e. does not hold at any world in play), holds with chance 0; and whatever is indeterminate (i.e. holds at some worlds in play but not at others), holds with some chance strictly between 0 and 1.

The idea underlying this framework may be traced to Leibniz's claim that chance is the 'degree of possibility' (Cirilo de Melo and Cussens, 2004). And this sort of framework is endorsed by those who accept the 'basic chance principle' (Bigelow, Collins, and Pargetter, 1993), which entails that if the chance of some claim A is greater than zero, then that claim A must be true in at least one possible world that 'matches' the actual world up until time t. This claim is accepted by many philosophers, as described in (Salmon, 2019)—though Salmon himself rejects the basic chance principle and the popular framework that I have described. Whether this framework is the right way to think about chance is up for debate—but for the rest of this chapter I will simply use the framework without delving into this question.

This chance framework suggests one way of interpreting the probability framework described in chapter 3. Recall that a probability space is a mathematical object, consisting of a set of states, an algebra over that set, and a function from that algebra into the range of numbers from 0 to 1. Such a probability space can be interpreted in various ways. As we have seen,

---

[13] More generally, philosophers define what it is for something to hold determinately *at a world* and *at a time*: here, as a sort of shorthand, I just talk about something just holding determinately— by which I mean that it holds determinately at the actual world at the current time.

THE PRINCIPAL PRINCIPLE 93

on one such interpretation a probability space can be used to represent an agent's epistemic state at a time, with the function interpreted as the credence function of the agent. On another interpretation, a probability space can be used to represent chance (at a time). On this interpretation— assuming the popular view of chance described in this section—we can interpret states as metaphysically possible worlds, with the chance function mapping sets of these states onto numbers between 0 and 1.

With this framework in mind, then, we can consider whether chance claims are hyperintensional or merely intensional. To crystallize this issue, let's return to our lottery example, and suppose that Alice has two names: she has the name 'Alice' and she also has the name 'Ms Smith'. The chance that Alice will win the lottery is 1/26. What is the chance that Ms Smith will win the lottery? Could it be other than 1/26? It seems that it cannot. For suppose (for the sake of argument) that the chance that Ms Smith will win the lottery is greater than the chance that Alice will win the lottery. Then there must be a chance greater than zero that Ms Smith will win the lottery and Alice will not. And in that case there must be some world in play at which Ms Smith wins the lottery and Alice does not. But there can be no such world—for there is no world at which Ms Smith is not Alice: 'Ms Smith' and 'Alice' are both rigid designators, and so pick out exactly the same person at every possible world, and so at any world where Ms Smith wins the lottery, Alice wins the lottery. Thus it cannot be that the chance that Ms Smith wins the lottery is greater than the chance that Alice wins the lottery, and for similar reasons it cannot be that the chance that Alice wins the lottery is greater than the chance that Ms Smith wins the lottery, and so the chance that Alice wins the lottery must be exactly the same as the chance that Ms Smith wins the lottery. More generally, then, given the framework that we are working with, within chance claims we can safely substitute co-referring names: chance claims are intensional but not hyperintensional.[14]

With this clarified, let us turn to a problem for the Principal Principle: the problem of the contingent a priori.

---

[14] This might be thought to create a problem for the Principal Principle, for presumably the chance operator assigns values to sets of metaphysically possible worlds, while the credence operator (being hyperintensional) must assign values to some other sorts of objects—possibly sets of more fine-grained worlds (see chapter 8). But it is not obvious that there is an issue here: plausibly we can interpret the chance framework in such a way that the chance operator assigns values to the same sorts of objects as the credence operator—it's just that it will assign the value zero to any fine-grained worlds that are not metaphysically possible, and it will assign the same value to any pair of fine-grained worlds that correspond to the same metaphysically possible world.

94 IMPLICATIONS FOR RATIONALITY

5.3.3 The Principal Principle and the contingent a priori

The problem that I discuss below may seem to concern a rather obscure case, and you might wonder why it should be of much interest. The answer to this worry is that the Principal Principle itself is considered to be of great importance, and so a problem which requires a reformulation of the principle is itself significant. As stated previously, the notion of chance plays a major role in numerous disciplines, including the natural and social sciences, decision theory, and ethical debates. And the Principal Principle is so called because it is supposed to specify *the* essential role of chance: 'A feature of Reality deserves the name of chance to the extent that it occupies the definitive role of chance; and occupying the role means obeying the [Principal Principle]' (Lewis, 1994, p. 489).[15] Thus a change in the formulation of the Principal Principle has the potential for major repercussions—for our understanding of chance, and ultimately for the various fields that rely on that concept.

Let us turn, then, to the problem. We start by taking the lottery case above, and we coin the name 'Lucky' to refer to the person—whoever it is—who will win the lottery. What is the chance that Lucky will win the lottery? Well, you are certain (let us assume) that the chance that Alice will win the lottery is $\frac{1}{26}$, and that the chance that Bob will win the lottery is $\frac{1}{26}$, and so on. You are also certain that Lucky is one of Alice to Zebedee: either Lucky is Alice, or Lucky is Bob, or ... Lucky is Zebedee. Let's start by supposing that Lucky is Alice. In that case, 'Lucky' and 'Alice' have the same extension, and so (as they are both names) the same intension. Given that the chance context is not hyperintensional, we can substitute 'Lucky' for 'Alice' safely within the chance context. Thus as the chance that Alice will win the lottery is $\frac{1}{26}$, it follows (assuming that Lucky is Alice) that the chance that Lucky will win the lottery is $\frac{1}{26}$. Thus if Lucky is Alice, then the chance that Lucky will win the lottery is $\frac{1}{26}$. For similar reasons, if Lucky is Bob, then the chance that Lucky will win the lottery is $\frac{1}{26}$—and so on for each of Alice to Zebedee. As you are certain that Lucky is one of Alice to Zebedee, you can be certain that the chance that Lucky will win the lottery is $\frac{1}{26}$. According to the Principal Principle, you ought to defer to the chance function, and as in this case you are certain that the chance that Lucky will win the lottery

---

[15] The version of the Principal Principle that Lewis refers to here is the 'Old Principle', which is given below.

THE PRINCIPAL PRINCIPLE 95

is ½₆, the Principal Principle dictates that you ought to have a credence of ½₆ that Lucky will win the lottery. But that is clearly wrong. Your credence that Lucky will win the lottery is, and rationally ought to be, 1: if you are in any doubt about this, just consider how you would bet on the claim, or whether or not you would be surprised to later hear the news that Lucky had won.[16] Here, then, we seem to have a violation of the Principal Principle (Hawthorne and Lasonen-Aarnio, 2009).

I begin by considering two natural responses to this problem. A first natural response is to object that there is something dodgy about the way that the name 'Lucky' is coined. After all (the thought might go) the lottery has not yet been held, and it is currently indeterminate who will win. How, then, can we coin a name for the person who *will* win, when as yet it is still unsettled who that will be? This worry might have its roots in the idea that claims about the future are not presently either true or false unless their truth or falsity is already determined. I disagree with this: I think that it may be true now that Alice will win, even if that is not yet determined. I am persuaded by the thought that if I now say 'Alice will win', and then Alice does indeed go on to win, then I can reasonably say that my utterance (at this earlier time) was true. Thus I think that it is true now—before the lottery is drawn—that Alice will win, or else true now that Bob will win, or else...true now that Zebedee will win. And on this view, it seems legitimate to coin a name for the person who will win. Not everyone will find this view persuasive—but I note that it is a view that fits well with the framework that we are assuming. On this framework, each possible world is a complete history—a past, present, and future. At some worlds, Alice wins the lottery; at some worlds, Bob wins—and so on. 'Lucky' is just the name of the person who wins the lottery in the actual world.[17]

---

[16] There is an interesting question here over whether there is a reading on which your credence that Lucky won the lottery is ½₆ rather than 1. Might there be a *de re* reading? Might we say that though you are sure that Lucky will win, still there is a sense in which you are unsure *of Lucky* whether (s)he will be the winner? I am persuaded that there is no such reading, and more generally that there is no epistemic *de re* (Yalcin, 2015; Ninan, 2018).

[17] If you accept the chance framework described but still think that there is something dodgy about coining the name 'Lucky', then we can construct the example without using the name 'Lucky' at all, but instead using the definite description 'the person who actually wins the lottery'. Though 'the person who actually wins the lottery' is a definite description, it is a rigidified definite description, referring to the same person at every possible world. If in fact Alice goes on to win the lottery, then the rigidified definite description refers to her—and it picks her out at every possible world. In that case 'the person who actually wins the lottery' has the same intension as 'Alice', and so given that the chance that Alice wins the lottery is ½₆, the

96 IMPLICATIONS FOR RATIONALITY

A second natural response is to object that the chance of Lucky winning the lottery is not ½₆, but 1. But on the framework we are assuming (rightly or wrongly) it seems undeniable that the chance of Lucky winning the lottery is indeed ½₆. To see this, let's suppose for example that Alice will win the lottery (bearing in mind that we could run a parallel argument for any of the other ticket-holders), so Alice and Lucky are one and the same. If the chance of Lucky winning the lottery was higher than the chance of Alice winning, then there would have to be a possible world in play where Lucky wins the lottery and Alice does not, but there is no such possible world, for at every possible world where Lucky wins the lottery, Alice—being the same person—also wins the lottery. We could similarly show that the chance of Lucky winning the lottery cannot be lower than the chance of Alice winning. Thus the chance that Lucky wins the lottery must be the same as the chance that Alice wins the lottery— namely ½₆. The chance function—unlike the credence function—does not create a hyperintensional context, and the chance that Lucky wins the lottery is the same as the chance that Alice wins the lottery (if 'Lucky' and 'Alice' happen to pick out the same person). The name 'Lucky' is importantly different here from the definite description 'the person who wins the lottery'. At least on one reading, the chance that the person who wins the lottery wins the lottery is 1: the person who wins the lottery wins the lottery at every possible world. In contrast, Lucky does not win the lottery at every possible world: the name 'Lucky' picks out the person—whoever it is— who wins the lottery in the actual world.

Thus the Principal Principle faces a problem. The claim 'Lucky wins' is knowable a priori—and so any rational agent's credence in the claim is 1— and yet there are possible worlds *still in play* where it is true, and possible worlds *still in play* where it is false—and so its chance is strictly between 0 and 1. The claim 'Lucky wins' is a particular sort of example of the contingent a priori, and there are a variety of other such claims which similarly create a problem for the Principal Principle. How should this problem be addressed? Hawthorne and Lasonen write: 'To say the least, the Principal Principle needs to be revised in the light of contingent *a priori* knowledge' (Hawthorne and Lasonen-Aarnio, 2009, p. 97). And several theorists have attempted such revisions or restrictions (Schulz, 2010; Schwartz, 2014; Gallow, 2021). I agree that some such revision is

---

chance that the person who actually wins the lottery wins the lottery is also ½₆. And yet of course your credence that the person who actually wins the lottery wins the lottery is 1. Thus we have a violation of the Principal Principle.

THE PRINCIPAL PRINCIPLE   97

needed, and here I put forward a proposed revision that I argue comes with a sound rationale.[18]

To explain my proposed revision, I begin by showing how the Principal Principle has already been revised in response to a different set of problems. To see these problems clearly, we need to make the Principal Principle more precise. Above I simply stated that the principle required rational agents to defer to the chance function—but here is a more precise version of Lewis's statement of the principle, which he calls 'The Old Principle' (Lewis, 1994):

Old Principle: $Cr(A|H_tT) = Ch_t(A)$

Here Cr is a rational agent's initial (prior) credence function—that is, her credence before gaining any evidence at all; Ch_t is the chance function at time t; A is any proposition; H_t states all the historical facts about the (actual) world up till time t; and T states all of the laws—including probabilistic laws. On Lewis's account, we can think of T as entailing all the true history-to-chance conditionals: so, for example, from facts about the way that a coin has been formed, and other facts about the world up till time t, it will follow from T that the chance of the coin landing heads on its next toss is ½. Thus from H_t and T it will follow that any given proposition A has some particular chance at time t.

Now that we have a more precise version of the Principal Principle, I turn to the problems that it faces. We have already seen that the principle faces a problem with certain cases of the contingent a priori: the current chance that Lucky wins is 1/26, and yet were you to conditionalize (just) on the entire history of the world and the laws, your credence that Lucky wins would be 1, for you know a priori that Lucky wins. This is the problem recently raised against the Principal Principle—but the principle also faces two other well-known problems.

5.3.4 Old problems for the Principal Principle

The first is the problem of crystal balls (Hall, 1994).[19] The Old Principle states that conditional on H_t and T, a rational agent's credence in A ought to

---

[18] One initially appealing response to this problem involves the machinery of two-dimensionalism, and this response is discussed though not ultimately endorsed in (Schulz, 2010, pp. 122–6). I argue in chapter 7 that the objects of credence are not primary intensions, and for this reason I would reject a two-dimensionalist response to the problem discussed here.

[19] Not all authors agree that this is a problem—see (Joyce, 2007; Meacham, 2010; Spencer, 2020).

98 IMPLICATIONS FOR RATIONALITY

equal the chance of A. If the agent were to conditionalize on information beyond H_t and T—information about what happens after t, perhaps—then the Principal Principle does not require that the agent's credence in A should equal the chance at t of A. Lewis classifies such information as 'inadmissible'. Defining exactly what inadmissibility amounts to has proved difficult. On first introducing the idea, Lewis stated that he had no definition of admissibility to offer, but confirmed that H_t and T were both admissible at t (Lewis, 1987, pp. 92–6). A problem arises, however, in cases where H_t and T convey direct information about the future. For example, suppose that a crystal ball has given accurate predictions many times before t, and that shortly before t it was consulted about the outcome of a future coin toss—a coin toss that will happen after t, and that will be a genuinely indeterministic process—and an image of the coin landing heads hovered within the ball. This information—about the past successes and the image hovering in the ball—is part of H_t, and conditional on this information, a rational agent may have a very high credence that (HEADS) the coin will land heads, even though the chance of HEADS remains at ½. This, then, is a counterexample to the Principal Principle.[20]

The second problem for the Principal Principle is the problem of undermining futures. To see the problem here, let's take a simple theory of chance, and a toy example. Suppose that the world contains just one coin, and it wears out after two tosses. You know that it has already been tossed once, and that it landed heads. It will be tossed just one more time in the future. That's our toy example. And now our simple theory of chance is that the chances are fixed by the actual frequencies across all of time: so the chance that this coin will land heads on any given toss is fixed by the proportion of times that it has or will land heads throughout all of its tosses—of which we know there are just two. What is your credence that the coin lands heads on its next toss (HEADS), conditional on H_t and T, bearing in mind that from H_t and T it follows (let's suppose) that the chance of HEADS is 0.5? Well, according to the Old Principle, your conditional credence in HEADS ought to be 0.5. But from the fact that the chance of HEADS is 0.5, you can infer that it must land tails on its next toss: only if it lands tails on its next toss will

---

[20] Here we might suppose that the problem raised in the previous section (concerning cases of the contingent a priori) is just a version of this problem of inadmissible evidence, for perhaps in coining or acquiring the name 'Lucky' we somehow gain inadmissible evidence. However there is no obvious reason to think that the evidence gained is inadmissible: for example, we might acquire the name 'Lucky' by watching a dubbing event take place—but it's hard to see how observing the dubbing event could give us inadmissible evidence (Nolan, 2016, p. 302).

the proportion of heads tosses to tails tosses throughout all time be 0.5. A future in which it lands heads is an 'undermining future'—in that it would undermine the claim that the chance of its landing heads is 0.5. Thus conditional on H_t and T, your credence in HEADS should be 0—in violation of the Principal Principle.

Here I have stated the problem using a toy example and a simple theory of chance, but the problem can also affect more realistic examples and more complex theories. In particular, the problem holds if we replace our simple theory of chance (on which chances are fixed by the actual frequencies across all of time) with any 'Humean Supervenience' theory, on which facts about chance are determined by non-probabilistic matters of fact. To see this, suppose that our theory of chance, T, is equivalent to some non-probabilistic matter of fact, F: this might be, for example, a fact about the frequency of all types of event across all time, or it may be something more complex. H_t and T together will entail that various propositions have various chances at time t. And in particular they will entail that proposition F has some particular chance at t, which may be less than 1. In that case, the Principal Principle would require that your credence in F (conditional on H_t and T) should be less than 1. And yet the probability axioms require that your credence in F (conditional on H_t and T) should be 1, given that T is equivalent to F. This problem (the 'Big Bad Bug' (Lewis, 1986a)) is the second of the two well-known problems for the Principal Principle.

Several responses to these problems have been put forward in the literature (Lewis, 1994; Hall, 1994; Thau, 1994; Hoefer, 1997; Roberts, 2001; Ismael, 2008; Nolan, 2016). Here I focus on just one famous response (Hall, 1994; Lewis, 1994). The Principal Principle is a deference principle: it tells us to treat chance as an expert. But experts come in different varieties. In Hall's terminology, we can distinguish between 'database experts' and 'analyst experts' (Hall, 2004, p. 100). A database expert is more knowledgeable than you: she knows all you know, and more. If we assume further that our database expert is perfectly rational, then it seems that you should simply defer to her. An analyst expert, in contrast, may have more knowledge than you in some areas, and perhaps have other virtues besides, but may lack some knowledge that you have. You are not rationally required to defer to an analyst expert. Rather, you are rationally required to defer to the analyst expert's credence function conditionalized on the information that you have.

To see this, we can take an everyday example. Suppose that you visit your doctor because you have been suffering from a particular sort of pain

100 IMPLICATIONS FOR RATIONALITY

which you think is probably tendonitis. Your doctor is, you are sure, very knowledgeable about the prevalence of all sorts of conditions, their likelihood, and their symptoms, and you're sure that she is an excellent reasoner besides. As you first enter her surgery, should you defer to her over the claim that you have tendonitis? Surely not! First you need to describe your symptoms and history: that is, you need to give her the relevant information that you have. Once you have done so, *then* you should defer to her. You should defer to the credence function that she would have (or will have) when she has conditionalized on all of your evidence. Thus the doctor at the start of your appointment is an analyst expert rather than a database expert.

Hall (Hall, 1994) and Lewis (Lewis, 1994) argue that objective chance is similarly an analyst rather than database expert. You are not rationally required to defer to the chance function simpliciter, but rather to the chance function conditionalized on the evidence on which you are conditionalizing. With this in mind, Hall and Lewis construct an alternative to the Old Principle—the New Principle.

New Principle: $Cr(A|H_tT) = Ch_t(A|T)$

According to this principle you do not simply defer to the chance function, but rather defer to the chance function conditionalized on T—the probabilistic laws. And we can see the rationale for this shift: if we are considering what credence you should have *conditional on T*, then it makes sense to track the chance function *also conditional on T* (Hall, 1994, p. 511). You might wonder here why we don't also have the chance function conditionalize on H_t—the entire past history of the world—and the answer is simply that the chance function has already conditionalized on H_t: all historical facts get a chance of 1, or in other words $Ch_t(H_t) = 1$.

One great advantage of switching to this New Principle is that it can handle the problem of undermining futures. To see how, let's apply this New Principle to our toy example of the coin toss—replacing 'A' with 'HEADS'. On the right-hand side we now no longer simply have the chance of HEADS, but the chance of HEADS conditional on T. This is equal to the chance of HEADS conditional on T and H_t (given that the chance at t of H_t is 1). From H_t and T, it follows that the chance of HEADS is—let's say—0.5. And yet this is incompatible with HEADS: given our simple theory of chance, the chance of HEADS is 0.5 (given the history of the world) only if the next coin toss lands tails—i.e. only if HEADS is false. Thus HEADS is incompatible with T and H_t, and so the chance of HEADS conditional on

THE PRINCIPAL PRINCIPLE   101

these is 0. Here, then, the New Principle gives the right result: the chance of HEADS conditional on T is 0, and this is the same as your credence in HEADS conditional on H_t and T. Thus it seems that the New Principle can handle the problem of undermining futures. Arguably, it can also handle the problem of crystal ball cases, and has no need for the concept of admissible evidence—though this is a matter of debate (Hall, 1994; Strevens, 1995). I do not try to settle this debate here, but instead consider whether we can extend the thinking behind the New Principle to handle the contingent a priori cases.

5.3.5 The New New Principle

There is a sound rationale in support of the New Principle: we should not simply defer to the chance function, but rather to the chance function conditionalized on all the information on which we are conditionalizing. It seems that we can similarly apply this rationale in the case of the contingent a priori. We know, of course, that Lucky wins: we are not exactly *conditionalizing* on this, but it is a claim that we know nonetheless. We should, then, not defer simply to the chance function, but to the chance function conditionalized on this claim. We can generalize this idea. Every a priori claim is a claim that we already know. Whenever a claim is a priori and contingent, the chance function may not have conditionalized on it: specifically, those claims which are a priori but not yet determinate will be claims that we know, but to which chance assigns a credence less than 1. We should defer to the chance function conditionalized on all such claims. For simplicity, we can just say that we should defer to the chance function conditionalized on all a priori claims. Let us say that X is the conjunction of all a priori claims— including both those that are contingent and those that are necessary. Then we can write the amended Principal Principle as follows:

The New New Principle: $Cr(A|H_tT) = Ch_t(A|TX)$

This is my proposed amendment to the Principal Principle. It is a natural extension of the New Principle, and it has a good motivation—similar to the motivation that prompted the development of the New Principle in the first place. The New New Principle can handle all the problematic cases of the contingent a priori. For example, the principle does not require you to set your credence that Lucky wins equal to $1/26$, but rather equal to

102 IMPLICATIONS FOR RATIONALITY

1—for the chance that Lucky wins, conditional on the claim that Lucky wins, is 1.

But surely there is a problem with this New New Principle? To see the problem here, suppose that, as it happens, Alice is going to win the lottery (although that is not yet determined). In that case, Alice is Lucky. And this identity claim holds across all possible worlds, and so the chance function must assign a value of 1 to the claim that Alice is Lucky—even before the lottery is drawn. Now, as explained in the last paragraph, the chance function—conditionalized on all a priori truths—assigns a value of 1 to the claim that Lucky wins. Thus it must also assign a value of 1 to the claim that Alice wins, even before the lottery has been drawn. This result can be generalized: if Bob is the person who will win, then the chance function—conditionalized on all a priori truths—must assign a value of 1 to the claim that that Bob wins, and so on for all other candidates. And more generally still: the chance function conditionalized on all a priori truths assigns a value of 1 to all truths and 0 to all falsehoods—whether these are already determined or not. A quick way to see how we can generalize the result in this way involves the actually operator. Take any claim P—where P might be determinately true, determinately false, or currently indeterminate. Whether P is currently determinate or not, the claim *actually P* is determinate: if P is true, then *actually P* is true in all metaphysically possible worlds, and so has a chance of 1; and similarly if P is false, then *actually P* has a chance of 0. Thus should P be true, the chance function assigns a value of 1 to *actually P*, and should P be false, the chance function assigns a value of 0 to *actually P*. Now, it is a priori that P *iff actually P*, and so the chance function—conditionalized on all a priori truths—assigns the same value to P as it assigns to *actually P*. In short, the chance function—conditionalized on all a priori truths—is omniscient: it assigns a value of 1 to any true claim and 0 to any false claim. Thus it seems that we might as well replace the right-hand side of the New New Principle with a simple truth-value function, that assigns 1 to any truth and 0 to any falsehood. Doesn't this principle, then, require rational agents to be omniscient?

In fact the principle does not require omniscience, for on the left-hand side of the principle we do not simply have a credence function, but a credence function conditionalized on H_t and T, and from H_t and T the chances (at t) of all events follow. Given the chance framework that we are assuming, it is unsurprising that an agent who is conditionalizing in this way would be omniscient. The chance function assigns 1 to all necessary truths, and a rational agent assigns 1 to all a priori truths—and from all necessary and a priori truths, all truths can be inferred. To see this recall that for any true claim P, the claim *actually P* is a necessary truth; and (whether P is true or not) *P iff actually P* holds a priori; and from *actually P* and *P iff actually P*, the claim P follows—and so more generally from all necessary and a priori truths, every true claim P follows. Thus an agent conditionalizing on H_t and T ought indeed to be omniscient. It does not follow that ordinary agents ought to be omniscient, for ordinary agents do not know the chances of all claims: for example, a typical agent at t—before the lottery has been drawn— would not know that the chance that Alice is Lucky is 1; and a typical agent would not know, when P is still indeterminate, whether the chance of *actually P* is 1 or 0. The New New Principle does not, then, make unreasonable demands on an ordinary agent.

But, then, we might wonder—what use is this New New Principle? The thought that lay behind the original Principal Principle was that our credences should be guided by the chances: as far as she is able, an agent should match her credence in a claim to the value that chance assigns to that claim. And this was thought to place an important constraint on what chance must be: a feature of reality can count as chance only if it fulfils the role set out for it in the Principal Principle. But now that we have arrived at the New New Principle, we find that it is rather different from the principle that we originally had in mind. Let us re-examine it with an eye to what constraints—if any—it places on the chance operator:

The New New Principle: $Cr(A|H_tT) = Ch_t(A|TX)$

On the left-hand side, we have—effectively—the truth-value operator: that is, an operator that assigns 1 to A iff A is true and 0 to A iff A is false. Let us substitute this truth-value operator (symbolized by 'V') to give the following New New* Principle:

The New New* Principle: $V(A) = Ch_t(A|TX)$

We can see, then, that the constraint on chance is simply this: the chance of any proposition, conditional on the conjunction of T and X, must be 1 if the proposition is true and 0 if the proposition is false. One way that a chance operator Ch_t can meet this requirement is by assigning 1 to every necessarily true proposition, and 0 to every necessarily false proposition. For then Ch_t will assign 1 to *actually P* whenever P is true, and 0 to *actually P* whenever P is false, and so conditionalized on all a priori truths (including the claim

104 IMPLICATIONS FOR RATIONALITY

that *actually P iff P*) Ch_t will assign 1 to P whenever P is true, and 0 to P whenever P is false. Provided, then, that an operator assigns 1 to all necessary truths and 0 to all necessary falsehoods, it can play the role that the New New* Principle sets out, and so (in this respect at least) it is worthy to be called the chance operator. Thus the constraint that this principle places on chance turns out to be rather thin, and not, as Lewis claimed, 'rich in consequences that are central to our ordinary ways of thinking about chance' (Lewis, 1987, p. 288).

This unravelling of the Principal Principle is a symptom of the deep difference between chance and credence claims. Chance claims are intensional, but not hyperintensional: all necessarily equivalent propositions are assigned the same chance, and necessary propositions are all assigned the value 1. In contrast, credence claims are hyperintensional: propositions that are necessarily equivalent can be assigned different credences, and necessary propositions need not be assigned the value 1. But the credence operator will assign the same value to all propositions that are a priori equivalent, and a value of 1 to any proposition that can be known to be true a priori.[21] Given these fundamental differences between credence and chance claims, it is unsurprising that there are difficulties and complications in a principle that connects them.

5.4 Chapter summary

In this part of the book, I am exploring some of the implications of my tenet that credence claims are opaque. In this chapter I have focused on two principles of rationality: the Reflection Principle and the Principal Principle. Both principles are affected by the fact that credence claims are opaque, and we can trace a range of problems and counterexamples to these principles to this source.

In the next chapter I turn to consider some of the practical implications of the tenet.

---

[21] Throughout this section I have assumed that any given credence function—at least, any credence function of a rational person—assigns the same value to all propositions that are a priori equivalent, and a value of 1 to any proposition that is knowable a priori. There are reasons to doubt this assumption, but these doubts won't help us here: that we are reasonably certain of a good many a priori truths (including that *actually A iff A*) is enough to create difficulties for the Principal Principle.

6
Practical Implications

6.1 Introduction

In this part of the book I am arguing that the fact that credence claims are opaque is important: it has numerous implications that theorists working with the Bayesian framework need to be aware of. In the last chapter (5) I focused on theoretical implications—and in particular implications for principles of rationality. In this chapter (6) I turn to practical implications. I begin by introducing decision theory, and then explore two cases where we can be led astray if we do not keep firmly in mind that credence claims are opaque: the first is the two-envelope paradox, and the second concerns the idea—widely discussed in welfare economics—of prospects for individuals.

I begin with a brief introduction to decision theory.

6.2 Decision theory

Decision theorists are interested in what people do—and what people should do—when deciding how to act. Here is an example: let's suppose that you are walking to meet a friend, and are faced with a choice between taking a shortcut across a field and going on a path around the edge. How should you go about deciding what to do? Let's suppose that you are quite certain how things will turn out under each option: if you take the shortcut, then you'll arrive on time but muddy, whereas if you take the path around the edge then you will be slightly late but not muddy, and there are no other relevant differences between the outcomes (e.g. you will enjoy both walks equally well). All you have to do, then, is decide which outcome is best, and this will dictate what you ought to do. Here, of course, we might question what exactly is meant by 'best' and what is meant by 'ought': on one reading, what is 'best' is what is morally best, and so the corresponding option is the one that you morally ought to choose, whereas on another reading what is 'best' is simply whatever you all-things-consider prefer, and the corresponding option is the one that it would be practically rational for you to choose—and

The Objects of Credence. Anna Mahtani, Oxford University Press. © Anna Mahtani 2024.
DOI: 10.1093/oso/9780198847892.003.0006

106 PRACTICAL IMPLICATIONS

Table 6.1

	Event e_1: The field is muddy $Cr(e_1) = 0.5$	Event e_2: The field is not muddy $Cr(e_2) = 0.5$
Cross the field	Late and muddy	On time and not muddy
Go round by the path	Late and not muddy	Late and not muddy

other readings are possible too. I will return to these issues later in the chapter, but to begin with let's assume a focus on what you prefer, and so on what option it is practically rational for you to choose.

In the case I described above you knew how each choice would turn out, but in many cases we are uncertain what the outcome of our choices will be. For example, let's change the above example by supposing that the field is sometimes almost impassably muddy, and then going across the field would not only get you muddy but also be no quicker than taking the path round the edge, but at other times the field is very dry, in which case if you were to cross the field you would arrive on time and not muddy. Let's suppose that at the time when you are faced with the decision, your credence (Cr) in each of these possibilities (which we can call 'events') is 0.5. We can represent your choice here using the decision table in Table 6.1.

We can suppose that your preferred outcome is to be on time and not muddy; the next best outcome is to be late and not muddy; and the worst-case scenario is to be late and muddy. Thus going round the path is the safe option as it is guaranteed to lead to your second-best outcome, whereas crossing the field is risky as it might lead to either your best outcome or your worst outcome. What, then, should you do? Plausibly that will depend not just on how you would order the outcomes, but also on how much you prefer one to another. For example, suppose that though you care a little bit about whether or not you arrive on time, what really matters to you is not getting muddy: in that case, you would choose to go round by the path. Alternatively, we can instead suppose that the most important thing to you is to be on time, and that you think there is hardly any point in turning up at all if you're going to be late—though if you are going to be late, then it is marginally better not to be muddy as well: in this case, you would choose to cross the field. Thus to predict how you would—and should—act, we need more than a mere ordering of preferences over the outcomes. That is, rather

DECISION THEORY 107

than merely placing the different outcomes on an ordinal scale (which only allows us to read off the order in which the outcomes are ranked), it seems necessary that the outcomes be placed on an interval scale (which would also allow us to see the relative differences in desirability between the outcomes), or even on a ratio scale (which would also include a zero point—allowing us to weigh the relative value of each option). Here there is some complexity over how such measurements could be obtained, and how they should be interpreted. For the purposes of this chapter, I will just assume without argument that each outcome can be assigned a number that captures the value (the 'utility') for the agent of that outcome. Let us suppose that in our example the numbers assigned to the outcomes are as in Table 6.2.

Table 6.2

	Event e_1: The field is muddy $Cr(e_1) = 0.5$	Event e_2: The field is not muddy $Cr(e_2) = 0.5$
Cross the field	Late and muddy: 2	On time and not muddy: 10
Go round by the path	Late and not muddy: 7	Late and not muddy: 7

Now that we have values assigned to each outcome, what does that tell us about how our agent does or should act? A popular view is that a rational agent maximizes expected utility (MEU): an agent is rationally required to perform the action which has the highest expected utility—or one such action, should there be more than one action which ties for first place. The expected utility of an action is calculated by summing the utilities of the action at each possible event, weighted by the credence that each event will obtain. As we can see in Table 6.3, in our example MEU requires the agent to go round by the path.

Table 6.3

	Event e_1: The field is muddy $Cr(e_1) = 0.5$	Event e_2: The field is not muddy $Cr(e_2) = 0.5$	Expected utility
Cross the field	Late and muddy: 2	On time and not muddy: 10	$(0.5*2) + (0.5*10) = 1 + 5 = 6$
Go round by the path	Late and not muddy: 7	Late and not muddy: 7	$(0.5*7) + (0.5*7) = 3.5 + 3.5 = 7$

108 PRACTICAL IMPLICATIONS

MEU is widely but not universally accepted as a rule for rational action. On some alternative views, it may be rational for an agent to be risk-avoiding or risk-seeking, and so pursue a different action from the one that maximizes expected utility (see (Buchak, 2013) for a recent version of such an account). And there are those who reject the claim that an agent's epistemic state can be modelled by a single credence function, and such theorists put forward alternative rules for rational action ((Bradley, 2017) discusses a range of such views). But for the purposes of this chapter we will assume that MEU is the right rule, and see what implications follow from the tenet (that credence claims are opaque) under this assumption: no doubt there would also be implications under other more complex decision rules too.

6.3 Decision theory and the objects of credence

The key tenet I have argued for in this book is that credence claims are opaque. How does that affect the decision theory framework that we are considering? Well, a first point to note is that in a decision table like that given in the last section, the top row of the table contains expressions each describing an event to which the agent assigns some positive credence. Given that credence claims are opaque, when describing these events it matters how the relevant objects are designated: we cannot substitute one name or definite description for another co-referring name or description without potentially changing the credence assigned.

To see an example of this, let's suppose that you are thinking of sending some flowers to your neighbour Edith, who you know to be currently unhappy, in the hope of cheering her up. You have also been conversing online with someone called 'Sneezy' who you know is allergic to flowers. You think it's possible that Edith is Sneezy (in which case the flowers will make her ill and fail to make her happy), but also possible that Edith isn't Sneezy (in which case the flowers will make her happy, and she'll remain well), so both of these possibilities need to appear in the first row of your decision table (Table 6.4).

Suppose that in fact Edith is Sneezy—so 'Edith' and 'Sneezy' co-refer. It does not follow that in the descriptions of events we can harmlessly substitute the name 'Sneezy' for the name 'Edith' or vice versa. You have a credence of 0.1 in event e_1—that *Edith is Sneezy*—but presumably you have a credence of 1 in the event that *Edith is Edith*. The descriptions of

DECISION THEORY AND THE OBJECTS OF CREDENCE   109

Table 6.4

	Event e_1: Edith is Sneezy $Cr(e_1) = 0.1$	Event e_2: Edith is not Sneezy $Cr(e_2) = 0.9$
Send flowers to Edith	Edith becomes ill and remains unhappy: 1	Edith remains well and becomes happy: 10
Do not send flowers to Edith	Edith remains well but unhappy: 7	Edith remains well but unhappy: 7

Table 6.5

	Event e_1: Edith is Sneezy $Cr(e_1) = 0.1$	Event e_2: Edith is not Sneezy $Cr(e_2) = 0.9$
Send the chocolates to Edith	Edith (and so Sneezy) is happy: 8	Edith is happy, but Sneezy is not: 7
Send the chocolates to Sneezy	Edith (and so Sneezy) is happy: 8	Sneezy is happy but Edith is not: 2

events that appear in the top row of a decision table appear there in an opaque context. This seems like something that our decision theory framework can accommodate. What about the other elements of the table—such as the outcomes and the actions? It seems clear that the descriptions of the outcomes also appear in an opaque context. To see this, let's alter the situation a bit and suppose that you are deciding whether to send a box of chocolates to Edith (by putting them through her letterbox) or to Sneezy (care of the business address that you can see on her website). You are quite sure, let's now suppose, that both Edith and Sneezy are currently sad but would be cheered up by a box of chocolates (Table 6.5).

According to this decision table, you value the outcome in which Edith is happy but Sneezy is not much more highly than you value the outcome in which Sneezy is happy but Edith is not. This is because, let's suppose, you care a lot more about Edith—your long-standing neighbour—than you do about Sneezy—a fleeting acquaintance online. This sounds plausible enough, but it entails that 'cares about' creates an opaque context: you can care about Edith a lot more than you care about Sneezy—even though Edith and Sneezy are (though you don't know it) the very same person. We can similarly say that 'prefers' creates an opaque context: you prefer Edith being happy over Sneezy being happy—even though Edith and Sneezy are one and

110   PRACTICAL IMPLICATIONS

the same. In short, just as descriptions of events in a decision table are opaque, so descriptions of outcomes are as well.

Finally let's consider how the descriptions of actions (in the leftmost column) should be understood. At first sight there might seem to be something very strange about the actions in table 6.5. For sending the chocolates to Edith is, in a sense, the same thing as sending the chocolates to Sneezy, given that (though you don't know it) Edith is Sneezy. How can the same action appear twice, with different outcomes under the same events? The answer is that these are two different actions. The actions listed in a decision table are those that you can make true at will: if you decide to carry out one of these actions, then you should be in no doubt as to how to go about it, or that you will succeed in carrying out the action so described. Thus in our table we have as an action 'send the chocolates to Edith', and not 'make Edith happy': to accomplish the former you know exactly what to do, but there are various ways that you might go about trying to accomplish the latter, with uncertain results. And 'send the chocolates to Edith' here names the sort of thing you might do by putting the chocolates through your neighbour's letterbox, or leaving them on her doorstep, while in contrast 'send the chocolates to Sneezy' might involve, say, posting them to Sneezy's business address. Thus there is certainly a sense in which these two actions are different—even if they actually lead to a similar result. Relatedly, many theorists define an action as a function from events to outcomes. And because the descriptions of both events and outcomes are fine-grained, the outcome of sending the chocolates to Edith is not the same at every event as the outcome of sending the chocolates to Sneezy—and so by this definition the actions are different. Thus the descriptions of actions in a decision table are also opaque.[1]

It seems that the standard decision theory framework can accommodate the fact that credence claims are opaque—and relatedly that the descriptions of events, outcomes, and actions in a decision table all appear in an

---

[1] It does not follow that all talk of actions in every context is opaque. For example, if I state that you sent the chocolates to Edith, then from my statement, together with the claim that Edith is Sneezy, it may follow that you sent the chocolates to Sneezy. In general, it may be that when we are attributing actions to agents, we can describe these actions in a variety of different ways, in line with Davidson's view of actions: 'I flip the switch, turn on the light, and illuminate the room. Unbeknownst to me I also alert a prowler to the fact that I am home. Here I do not do four things, but only one, of which four descriptions have been given' (Davidson, 1963, p. 686). My claim here is just that *within a decision table*, the action-descriptions appear in an opaque context.

THE TWO-ENVELOPE PARADOX   111

opaque context.[2] But these facts are important, and we need to keep them in mind when working with decision tables, as they have some wide-reaching implications. In the next section I illustrate one such implication for the two-envelope paradox.

6.4 The two-envelope paradox

Suppose that you have before you two envelopes, each of which contains a cheque for some amount of money. You have no idea how much money these two envelopes contain, but you do know that one contains twice as much money as the other. You select an envelope at random. But then you consider: should you stick with the envelope you've selected, or should you switch to the other one? There seems to be a good reason to switch, as you can see from Table 6.6, with 'M' standing for the amount of money in the envelope in your hand and 'N' standing for the amount of money in the other envelope.

Obviously the expected utility of sticking with your current envelope is M (you are guaranteed to get the amount that is in that envelope); whereas if you switch there is a 0.5 chance that you will get double M, and a 0.5 chance that you will get half of M, and so the expected utility of switching is $(0.5)(2M) + (0.5)(0.5M) = 1.25M$. Switching, then, seems to have higher expected utility than sticking, and so MEU requires you to switch. But this is a very strange result! Intuitively there is no rational reason for you to switch. And it is not just intuition that speaks against this result, for we can construct a parallel argument in favour of sticking, with the outcomes stated in terms of N (the amount of money in the other envelope), as shown in Table 6.7.

Table 6.6

	Event e_1: N = 2M Cr(e_1) = 0.5	Event e_2: N = 0.5M Cr(e_2) = 0.5	Expected Utility
Stick	M	M	M
Switch	2M	0.5M	1.25M

[2] The claim that the descriptions of events, outcomes, and actions in a decision table all appear in an opaque context fits well with Richard Jeffrey's view on which these are all propositions (Jeffrey, 1983).

112 PRACTICAL IMPLICATIONS

Table 6.7

	Event e_1: N = 2M $Cr(e_1) = 0.5$	Event e_2: N = 0.5M $Cr(e_2) = 0.5$	Expected Utility
Stick	0.5N	2N	1.25N
Switch	N	N	N

Table 6.8

	Event e_1: M = ⅓Z N = ⅔Z $Cr(e_1) = 0.5$	Event e_2: M = ⅔Z N = ⅓Z $Cr(e_2) = 0.5$	Expected Utility
Stick	⅓Z	⅔Z	0.5Z
Switch	⅔Z	⅓Z	0.5Z

What has gone wrong here? A first thought might be that this problem can be resolved by setting out the decision table differently. For example, we might let 'Z' stand for the amount of money in the two envelopes combined, and then set the decision problem out using Table 6.8.

Table 6.8 gives us the answer that we intuitively expect—that it is rationally permitted to either switch or stick, for both actions have the same expected utility. But the fact that we can construct a decision table that gives us the right result does not resolve the problem, for we still need to explain what was wrong with the previous two decision tables. At first sight, at least, they seemed to be perfectly in order, and it was only because we knew that they gave us the wrong results that we were alerted to their being somehow defective. In cases where we do not already know what result to expect, how can we spot decision tables that are defective in a similar way? In short, what rule can we give for constructing decision tables that guarantees that we won't construct a decision table that is defective, and what is the rationale for any such rule? A satisfactory answer to this question is needed to resolve the paradox. Here I do not attempt to solve the paradox in full generality: many versions of the paradox have been put forward, and there is no widespread agreement over the correct solution. My aim here is just to show how at least one version of the paradox connects with the key tenet of

THE TWO-ENVELOPE PARADOX   113

this book—that credence claims are opaque. I begin, then, by giving the specific version of the paradox that I am addressing here.

6.4.1 A specific version of the two-envelope paradox

In the set-up described above, I said that the envelopes were both known to contain some amount of money, and I put in place no restriction on the possible amounts that they might contain. But in any realistic version of this scenario there will be some upper bound (for the envelope can't contain more money than there is in the world), and some lower bound (for the envelope can't contain an amount of money smaller than the smallest unit of currency), and the range of possible amounts are discrete rather than continuous (Jackson, Menzies, and Oppy, 1994). It might be thought that these sorts of considerations are relevant to the decision to switch or stick, and should somehow be represented in the decision table. With this in mind, I focus on a simplified version, adapted from Terry Horgan's 'urn case' (Horgan, 2002). In this version there are just four possibilities: it might be that the envelope you have chosen contains £1 and the other contains £2 (1,2), or vice versa (2,1), or it might be that the envelope you have chosen contains £2 and the other £4 (2,4), or vice versa (4,2). Let us assume that your credences are divided equally between these four possibilities. On this more precise version of the problem, we can see that it is still possible to construct a decision table (see Table 6.9) on which it seems rationally required to switch (and a parallel table could be constructed to show in contrast that it is rationally required to stick).

Something has clearly gone wrong here—but what? The problem seems to be that the outcomes are stated in terms of 'M'. What sort of term is 'M'

Table 6.9

	Event e_1: 1,2 $Cr(e_1) = 0.25$	Event e_2: 2,1 $Cr(e_2) = 0.25$	Event e_3: 2,4 $Cr(e_3) = 0.25$	Event e_4: 4,2 $Cr(e_4) = 0.25$	Expected Utility
Stick	M	M	M	M	M
Switch	2M	0.5M	2M	0.5M	1.25M

114 PRACTICAL IMPLICATIONS

exactly, and can it play the role that is here being expected of it? We introduced 'M' to stand for the amount of money in your hand—but in what sense does 'M' stand for this amount?

6.4.2 'M' as a transparent designator

One view is that in this decision table 'M' works as a purely transparent—or 'referential'—term, standing for the amount actually in the envelope that you are holding, in which case we can harmlessly substitute any other term similarly designating that amount. Suppose, for example, that (though you don't know it) the amount actually in the envelope is £1, so M is £1. In that case (on the view we're considering) we can substitute '£1' for 'M' throughout.[3] But then the decision table above is obviously incorrect: for example, in event e_2 the outcome of sticking is £2 rather than £1, and so the 'M' placed in this cell must be wrong if it means the same as '£1'. This approach would be mandated if the decision table was an extensional context: that is, if the expressions within a decision table designating the various parts (the events, the actions, and the outcomes) could be safely substituted with any other co-referring designators. But I have argued that all of these designators (including the designators for the outcomes) are given in an opaque context, and so there is no reason to assume that 'M' functions as a purely referential term—particularly when the result is that a decision table which seemed compelling is shown by this approach to be so riddled with obvious errors.

---

[3] A variation on this approach seems to underlie the argument of (Jackson, Menzies, and Oppy, 1994). On their view, if you have an equal credence that the amount in the other envelope is 0.5M or 2M, then that is because for any amount of money that you think might be in the envelope you are holding, you have an equal credence that the amount of money in the other envelope is half or double this amount. And we can see that this is clearly not the case in our simple version: you think that the envelope that you are holding might contain just £1, but conditional on that being the case you do not have an equal credence that the amount of money in the other envelope is £0.50 or £2. In (Jackson, Menzies, and Oppy, 1994) the authors argue that you are thus in error to have an equal credence that the amount in the other envelope is 0.5M or 2M—for the claim that they take to motivate this assignment of equal credence has been shown to be false. But their argument here is too hasty: in your ignorance as to the value of M, it is indeed rationally justified to have an equal credence that the amount in the other envelope is 0.5M or 2M—as can be seen from Table 6.9. Thus their diagnosis of why you have this credence must be wrong. See (Horgan, 2002) and (Katz and Olin, 2007) for further discussion of this point.

6.4.3 'M' as a definite description

On a more charitable reading, 'M' is a definite description, and as definite descriptions do not always designate rigidly, perhaps 'M' is such a non-rigid designator. In that case it designates £1 when it appears in the column for event e_1, £2 when it appears in the column for event e_2, and so on. On this reading the decision table is not *obviously* incorrect—but is it acceptable to state outcomes in terms of such non-rigid definite descriptions? It is certainly not always acceptable to do so. To see this, consider the following scenario. Suppose that I have to choose between two gifts—a gift of £100 and a gift of £1000. Assume that I would prefer £1000 to £100, and that there are no other relevant differences between the two options, and then it seems clear that I am rationally required to select the gift of £1000. But now I can reason as follows: if I accept £1000, then I will get the amount that I am given (call it G); and if I accept £100, then I will get the amount that I am given (G). So in either case, the outcome is the same, and I am rationally permitted to carry out either action (Table 6.10).

Something has clearly gone wrong here, and the problem is that the outcomes are stated in terms of 'G', which stands for a non-rigid definite description, 'the amount that I am given'. What quantity 'G' denotes—and so how much value 'G' has for me—varies depending on which action I decide to carry out: I prefer the 'G' that appears in the second row of the table to the 'G' that appears in the third row of the table. As the value of 'G' changes in this way between the rows of the decision table, the fact that both actions have a guaranteed outcome of 'G' does not tell us that they are both rationally permissible. So if we are going to use decision tables to compare outcomes across different actions, then it seems we need the outcomes to be stated in terms that express the same thing wherever they appear.

The example involving 'G' shows us that it is important that we use terms that refer to the same quantity in each row in which they appear. And the two-envelope paradox seems to show that it is similarly important that

Table 6.10

	Outcome
Take the gift of £1000	G
Take the gift of £100	G

116 PRACTICAL IMPLICATIONS

Table 6.11

	Event e_1: 1 $Cr(e_1) = 0.5$	Event e_2: 10 $Cr(e_2) = 0.5$	Expected Utility
Stick	1	10	5.5
Doubled if £1	2	10	6
Doubled if £10	1	20	10.5

we use terms that refer to the same quantity in each column in which they appear. To see intuitively why this matters, we can start by thinking about a rather different case as follows.[4] You are handed an envelope that you are told contains either £1 or £10 (and your credence is divided equally between these possibilities), and you are given three options: you can choose to keep what's in the envelope; or you can choose a deal whereby it gets doubled iff it is £1; or you can choose a deal whereby it gets doubled iff it is £10. It seems clear enough that you should choose the deal whereby it gets doubled iff it is £10 (Table 6.11).

But now let's set out those choices again, using 'M' to stand for a non-rigid definite description denoting whatever is currently in your envelope (Table 6.12). Now it appears that the last two options ('doubled if £1' and 'doubled if £10') have equally good expected outcomes, and so are both rationally permissible. The problem here is that by stating the outcomes throughout as multiples of 'M', we obscure the fact that the value of M is higher at event e_2 than it is at event e_1, and so obscure the fact that it is of more value to double M at event e_2 than to double M at event e_1. Really we should be paying more attention to what happens at events where the value of M is high than to what happens at events where the value of M is low— but the expected utilities in the decision table do not reflect this. The same phenomenon is at work in the simple version of the two-envelope paradox described earlier: Table 6.13 is a decision table for a still simpler version of that problem to bring out the connection.

From the decision table, it looks as though the doubling of the outcome at event e_1 *more than* makes up for the halving of the outcome at event e_2: by doubling M, you get an extra M, whereas by halving M you lose just half an M. But in fact the doubling of the outcome at event e_1 *only just* makes up for the halving of the amount at event e_2, because what happens at event e_2 is more important given that M has a higher value there. Thus we can see why

---

[4] This is a simplification of a case discussed in (Schwitzgebel and Denver, 2008).

Table 6.12

	Event e_1: 1 $Cr(e_1) = 0.5$	Event e_2: 10 $Cr(e_2) = 0.5$	Expected Utility
Stick	M	M	M
Doubled if £1	2M	M	1.5M
Doubled if £10	M	2M	1.5M

Table 6.13

	Event e_1: 1,2 $Cr(e_1) = 0.5$	Event e_2: 2,1 $Cr(e_2) = 0.5$	Expected Utility
Stick	M	M	M
Switch	2M	0.5M	1.25M

it is problematic to express outcomes in terms that denote different quantities at different events.

To solve the two-envelope paradox, we need to give a rule for constructing decision tables: the rule needs to have a good rationale, and it needs to guarantee that our decision tables are not defective in the way that gave rise to the paradox. With this in mind, a good rule seems to be: ensure that your outcomes are stated in terms that have the same denotation across all events and actions—or at least, that denote things of the same value to you. This rule has a good rationale, as we have seen, and plausibly the two-envelope paradox arose because of a violation of this rule. To figure out the implications of this rule we can consider which sorts of designators are permitted when stating outcomes. We have seen that definite descriptions can create problems, but is it okay if they are rigidified? And are proper names—also being rigid designators—safe to use?

6.4.4 Are rigid designators safe to use as outcomes in decision tables?

Here it is important to bear in mind that credence claims are opaque—and indeed hyperintensional. The relevance of this point to the two-envelope paradox is compellingly made by (Horgan, 2002; Horgan, 2016), and here I explain his important point (albeit in a somewhat different way).

118 PRACTICAL IMPLICATIONS

Let's consider again the designator 'M' as it is used to label a possible outcome. If we understand 'M' as a definite description, meaning *the amount in the envelope I've selected*, then the outcome that it designates may vary across the different events described in the columns of a decision table—just as it varies across different metaphysically possible worlds. If instead we understand 'M' as a rigidified definite description, meaning *the amount that is actually in the envelope that I've selected*—what happens? Now the amount that it designates no longer varies across metaphysically possible worlds, for we have made 'M' into a rigid designator. But the amount that it designates may still vary across the events in a decision table. Take for example the following two events: the event at which the amounts *actually* in the envelope I've selected and the other envelope are (1,2) respectively; and the event at which the amounts *actually* in the envelope I've selected and the other envelope are (2,1) respectively. At the first of these events, 'M' designates £1, whereas at the second of these events, 'M' designates £2. Thus, even though we are now treating 'M' as a rigid designator, its value can still vary across events.

You might object that we should not have, in our decision table, descriptions of events involving the word 'actually'. But why not? Just as you are uncertain what is in the envelope you are holding, so you are similarly uncertain what is *actually* in the envelope that you are holding. Adding 'actually' makes no difference as far as your uncertainty goes. Adding 'actually' can make a difference where modal operators are concerned: for example, at a given metaphysically possible world, it might be that *necessarily P* does not hold, but *necessarily actually P* does. But the operator 'actually' typically does not make a different where epistemic operators are concerned: if you don't know whether P, then you don't know whether actually P either; and if you have a credence of v that P, then you'll also have a credence of v that actually P. Thus in Table 6.13, we can interpret the descriptions of events e_1 and e_2 as describing how things *actually* are, and the credences assigned to each event will still apply. On this interpretation of Table 6.13, the value of 'M'—even when understood as a rigidified designator—varies from event to event. Thus, even with 'M' understood as a rigid designator, we can still create a defective decision table.

6.4.5 A better restriction

It is not enough, then (and it is unnecessary), to ensure that we state outcomes in terms of designators that are rigid across metaphysically

THE TWO-ENVELOPE PARADOX   119

possible worlds.[5] Our requirement should rather be that we must state outcomes in terms that are rigid across the events listed in the decision table. As Horgan puts it, what we want here are 'epistemically rigid designators' (Horgan, 2002). And at first blush that would seem to mean that we must state them in terms such that we are certain what those terms denote. If we are uncertain what a given term denotes, then there might be one event at which it has one denotation, and another event at which it has a different denotation, and once again we have a defective decision table. But what does it mean for us to be certain *what a term denotes*? This is a slippery concept. Suppose that I pass you an envelope, and you have no idea what is inside. Do you know what 'the thing inside that envelope' denotes? In a sense you do: it denotes the thing that is inside the envelope! If I were to ask you what was inside the envelope, you'd be able to reply quite truthfully by saying: 'the thing inside the envelope'. You have a route to that object—a way of picking it out and referring to it: why isn't that enough to count as knowing what's inside? What else is needed? Do you need to also know what it looks like, or be able to pick it out by observation from a collection of similar objects? These requirements seem too stringent: if you know that my most recent drawing is inside the envelope, then surely you know what is inside even if you don't know what it looks like and couldn't pick it out of a group of similar objects. The very idea of *knowing what something is* seems to fall apart on examination, and plausibly what is required will vary with the context.[6] Fortunately, as far as outcomes are concerned, we don't need to try to unpick this expression, for what really matters is that we should be certain of the *value* of whatever the term denotes, such that at

---

[5] Compare (Katz and Olin, 2007), who write (using 'n' where I use 'M'): 'We are ... treating "n" as a rigid designator, that is, as a term that designates the same value in every possible world ... if we were not using "n" rigidly, we could not make sense of comparisons of utility across different possible states of the world, in which case our calculations of expected utility would not make sense' (Katz and Olin, 2007, p. 909). I claim that treating 'n' as a rigid designator in the standard sense—i.e. as rigid across metaphysically possible worlds—does not ensure that it refers to the same value across the different events (or 'states') of a decision table: there are events that are metaphysically possible but not epistemically possible (as (Katz and Olin, 2007; Katz and Olin, 2010) agree), but there are also events that are epistemically possible without being metaphysically possible.

[6] The questions here are connected with those sometimes raised against the idea of *de re* belief—and there is an interesting connection here, for (Katz and Olin, 2010) argue that the relevant credences in the decision table should be read as *de re* credences—and so that you do not in fact know that the envelope you are holding contains M—while (Sutton, 2010) disagrees and so objects to their proposed solution to the two-envelope paradox. I claim that the credences should not be read as *de re* credences, as my general claim that credence claims are opaque precludes this.

120 PRACTICAL IMPLICATIONS

every event the term denotes an object of the same value. That suggests the following restriction:

Sameness of value: An outcome in a decision table should be designated in such a way that the value of that outcome (so designated) is the same at every event—i.e. the agent should know with certainty the value of the outcome (so designated).

This rule guarantees that we will not construct decision tables that are defective in a way that gives rise to our simple version of the two-envelope paradox. The problem there was that we used the term 'M' to state the outcomes, without being certain of the value of 'M': we didn't know how much money was in the envelope we were holding, and so the value of 'M' varied across the different events. The proposed rule guides us safely away from such decision tables. But it is quite a demanding rule, and it can be softened.[7] To see why we might want to soften this rule, consider that in any realistic scenario, when we set out a decision table we do not spell out every source of uncertainty, but just those that are relevant for the problem at hand. Thus the events listed in the top row of a decision table are typically not complete states—that is, fully detailed ways that the world might be (for all we know). Rather they correspond to unions of such complete states. To see this, suppose that in our scenario you have a credence of 0.5 that the cheques inside the two envelopes are written in blue ink, and a credence of 0.5 that the cheques are written in black ink—and so each event listed in the decision table can be broken down further into more detailed possibilities: for example in Table 6.13 event e_1 includes the possibility at which there is £1 in the first envelope and £2 in the second and the cheques are written in black ink, and a possibility otherwise similar but with the cheques written in blue ink—and so on. But there is no need to have separate events spelling out this uncertainty in the decision table, with different outcomes under each ('a cheque for £2 written in black ink', 'a cheque for £2 written in blue ink', and so on). One reason why there is no need for that is that plausibly you don't care whether your cheque is written in blue ink or black ink: either way, its value to you is (let's suppose) exactly the value of £2, so we can just

---

[7] An alternative way to soften the rule would be to require not that the agent should know with certainty the value of the outcome, but simply that the agent should know with certainty that the value of the outcome is the same at every state. This requirement is close to that advocated by Horgan (personal correspondence).

THE TWO-ENVELOPE PARADOX   121

record the outcome as £2 with no need to have separate events specifying the colour of the ink. But suppose that you did have a preference for black ink over blue ink, so that to you the value of a cheque for £2 written in black ink was £2.01, and the value of a cheque written in blue ink was £1.99. Would we then need to increase the number of events in the table to capture this uncertainty and these differently valued outcomes? Or could we keep the events reasonably coarse, and describe the outcomes in a similarly coarse way (e.g. as 'a cheque for £2' rather than as 'a cheque for £2 written in blue ink', and so on)? If we keep the events and outcomes coarse in this way, then we won't be able to say that the terms in which the outcomes are given are such that you can be *certain* of the value of what they denote: if you were to learn that you were getting a cheque for £2, then you wouldn't be certain whether it would be worth £1.99 to you or £2.01. Nevertheless, if we assume that the colour of the ink is probabilistically independent (relative to your credence function) of the listed events, then we can say that at each event listed in the decision table, the *expectation* of the value of the object that 'a cheque for £2' denotes is £2. This suggests the following restriction on how outcomes should be designated in a decision table:

Sameness of expectation of value: An outcome in a decision table should be designated in such a way that under each event, the agent's *expectation* of the value of the outcome (so designated) is the same.

This rules out the defective decision tables that we have seen. For example, it rules out Table 6.13, for not only are you uncertain of the value of M as it appears in that table, but your expectation of that value is different at different events listed in the decision table: conditional on event e_1 obtaining, your expectation of the value of M is that of £1, but conditional on event e_2 obtaining, your expectation of the value of M is that of £2. That table and the other defective tables we have looked at are ruled out by the rule 'Sameness of expectation of value'.

We have arrived at the rule 'sameness of expectation of value' by reflecting on the fact that the expressions in a decision table denoting events, actions, and outcomes are in an opaque (or hyperintensional) context. To solve the two-envelope paradox, it is not enough (and it is unnecessary) to rule that the expressions for outcomes must be rigid across metaphysically possible worlds. For even if we use expressions that are rigid in this sense, they may not denote objects of the same value across all events. We need the expressions to be, as Horgan puts it, epistemically rigid. I hold that

122 PRACTICAL IMPLICATIONS

specifically what is required is that our expressions for the outcomes are such that the expected value of the object denoted is the same at every event. This position coheres well with others in the literature. For example (Schwitzgebel and Denver, 2008) arrive via a different route at a similar conclusion.[8]

It is worth noting here that though the two-envelope paradox is based on a contrived and unlikely scenario, there are plenty of realistic cases where similar issues arise, and the rule above can be of real practical help. To see this, let's begin by transferring the two-envelope case to a more realistic setting—while retaining the artificial simplicity of the example. Suppose that you are considering whether to part exchange your car. You are unsure of the value of your current car, but you have just been shown a potential alternative in a showroom. The salesman claims that the showroom car is around twice as valuable as your current car, but your friend thinks it is the other way round. You might be tempted to reason as follows: the value of my current car is C, and if I keep my current car obviously I will retain C; if I exchange my current car for the showroom car, then I might get 2C, or I might get ½C; with a credence of 0.5 in each of these possibilities, it seems that the expected value of switching is 1.25C, and so I should be willing to pay a quarter of the (expected) value of my current car in this part exchange. This reasoning obviously follows the defective reasoning in the two-envelope paradox, and we can now diagnose the error: in thinking through this decision problem, you should not express the outcomes in terms of C, for the expected value of C varies across the relevant events. Were you to learn that your car is worth twice as much as the showroom car (and nothing else at all), presumably your expectation of the value of your current car would increase; and it would similarly drop were you to learn that it is worth half as much as the showroom car. This example is of course still unrealistically simple, but by increasing the number of possible events and varying

---

[8] There is an interesting question over whether the principle I recommend ('Sameness of expectation of value') is assured by Leonard Savage's postulates on preferences (Savage, 1954). Savage relies on a number of postulates which together are designed to establish an expected utility representation of any agent who satisfies those postulates. One such postulate—P3— relates to my 'sameness of expectation of value' principle, but P3 (being an ordinal requirement) is not strong enough by itself to establish this principle. On Richard Bradley's view, Savage's choice of a cardinally (rather than merely ordinally) state-independent representation is 'entirely arbitrary' (Bradley, 2017, p. 61). Many thanks to Ze'ev Goldschmidt and Nicholas Makins for drawing my attention to this connection and for very valuable discussion on this point.

your credence in each, much more complex and realistic examples could be constructed—and the underlying point would still apply.

We might put the practical lesson from the two-envelope paradox as follows: in stating outcomes, always ensure that the currency in which you express those outcomes does not vary in expected value across states. For example, when considering the value of a car exchange, do not express the possible outcomes in terms of the value of your current car, if this varies across the relevant events. A natural thought is that you would be safer to express outcomes in terms of literal currency—such as pounds or dollars or similar. This may work well for many scenarios, but one difficulty is that you may not value money linearly: £2000 may not be worth twice as much to you as £1000. And another difficulty which can now be seen clearly is that in some scenarios the value of this currency can vary across the relevant events. We might, for example, be unsure whether to accept a bet on the pound dropping in value, and here our decision table may include events where the pound does drop in value and events where it does not. If we express the possible outcomes in terms of pounds, then we will find that the value of our currency varies across the events: should the pound drop in value, then obviously £100 will be worth less than if it does not. To avoid constructing a defective decision table here, we would need to switch to some currency whose expected value remains fixed across all relevant events.

6.4.6 Variations

Before moving on from the two-envelope paradox, I run through a few decision tables relating to the two-envelope paradox to check that our rule gives the right result in each case. First of all, let's take our initial decision table for the two-envelope paradox (Table 6.6).

I am assuming throughout that there is both an upper and lower bound to the amount of money that can be in either envelope. For simplicity, let's imagine that the envelopes might contain either £1 and £2, or £2 and £4— but the point holds for any finite range of values. Under event e_1 M is the smaller of the two amounts, and so might be either £1 or £2. Thus your expectation of the value of M at e_1 will be between the value of £1 and £2. Under event e_2 where M is the larger of the two amounts, your expectation of the value of M will be between the value of £2 and £4. Thus your expectation of the value of M is higher under e_2 than it is under e_1. As your expectation of the value of M varies across states, our rule is violated

124 PRACTICAL IMPLICATIONS

Table 6.6 (again)

	Event e_1: N = 2M $Cr(e_1) = 0.5$	Event e_2: N = 0.5M $Cr(e_2) = 0.5$	Expected Utility
Stick	M	M	M
Switch	2M	0.5M	1.25M

Table 6.8 (again)

	Event e_1: M = ⅓Z, N = ⅔Z $Cr(e_1) = 0.5$	Event e_2: M = ⅔Z, N = ⅓Z $Cr(e_2) = 0.5$	Expected Utility
Stick	⅓Z	⅔Z	0.5Z
Switch	⅔Z	⅓Z	0.5Z

and this decision table is classed as defective. In contrast, consider Table 6.8, constructed in terms of Z, the total amount in the two envelopes. Here of course (just as for M above) you are uncertain what value Z takes. But your expectation of Z's value is the same at event e_1 as at event e_2. Thus this decision table is not defective.

I turn now to consider a variation on the two-envelope paradox—the 'coin toss' version (Cargile, 1992). In this version, the envelope that you are holding is filled and then a coin is tossed to decide whether to put double that amount or half that amount into the other envelope. We can suppose that if (e_1) the coin lands heads, then double the amount in your envelope is put into the other envelope, and if (e_2) the coin lands tails, then half the amount in your envelope is put into the other envelope. For this version of the case, the table stated in terms of M is correct: your expectation of the amount in your own envelope is the same conditional on your being at either event e_1 or e_2. And here, the table constructed in terms of Z is defective: your expectation of Z is higher under the event in which the other envelope contains twice as much. In this case, then, it is indeed rational to switch—and this is intuitively the right result.[9]

---

[9] This coin toss case is different from 'peeking' cases (Clark and Shackel, 2000), where the cheques are already in the envelopes and you just get the chance to peek at what is inside. Once you have peeked inside your envelope, your expectation of the amount in your own envelope is

WELFARE ECONOMICS 125

In this section I have focused only on versions of the two-envelope paradox where we assume that there is an upper and lower limit to the amount of money that can be placed in the two envelopes, and that the range of possible amounts is discrete rather than continuous. Many think that the truly difficult versions of this paradox arise only in the cases where one or more of these assumptions is dropped, and there are many interesting theories and debates about the more difficult versions (Broome, 1995; Clark and Shackel, 2000; Chalmers, 2002; Meacham and Weisberg, 2003; Dietrich and List, 2005). My aim here has been to highlight the connection between this paradox and the fact that credence claims are opaque. It is because credence claims are opaque that we cannot solve the paradox by requiring that all outcomes be designated rigidly: it's no good ensuring that the terms denote objects with the same value across all metaphysically possible worlds. Rather, we must ensure that the terms denote objects with the same value—or at least the same expected value—across all the events listed in the decision table. This is one example of a case where the fact that credence claims are opaque is important for practical reasoning. In the next section I turn to a further implication, this time in welfare economics.

6.5 Welfare economics

So far in this chapter we have been looking at decisions from the perspective of a single decision maker, whose personal preferences fix the relevant utilities. The rules that we considered (e.g. MEU) were rules of practical rationality—i.e. rules that an agent ought to obey on pain of irrationality. I turn now to decisions that may affect outcomes for a group of people. These are the sorts of decisions that interest welfare economists, social choice theorists, and those working in distributive justice, and here the rules that are proposed may be classified as moral rather than rational.

Let's begin with a simple example. Suppose that you are a doctor, and you have two patients with suspected appendicitis: you have a credence of 0.8 that Alice (A) has appendicitis, and a credence of 0.8 that Belinda (B) has

---

indeed the same conditional on your being at either event, but Table 6.6 given in terms of M may still be incorrect. For—depending on what you see within the envelope—your credence in e_1 or e_2 may change: for example, in the simple case where there is an upper (£4) and lower (£1) limit to the amounts in the envelopes, if you see £4 in your own envelope, then $Cr(e_1)$ is 0 and $Cr(e_2)$ is 1. A similar result holds in more complex cases provided that there is an upper and lower bound to the amounts that can be in the envelopes.

126 PRACTICAL IMPLICATIONS

appendicitis, and these claims are probabilistically independent relative to your credence function.[10] And suppose that you have one dose of medicine that you know will cure appendicitis if wholly given to a single patient, or will slightly relieve the symptoms if split between two. Table 6.14 gives the utilities—or as we will call it here, the welfare[11]—for both A and B under each action and event. Having appendicitis and no medicine is a bad outcome (with a welfare value of 1); having appendicitis and getting a half dose of the medicine is not much better (with a welfare value of 2); having appendicitis and getting a full dose is a good outcome—as is not having appendicitis in the first place, regardless of whether you get the medicine (10).

Table 6.14

	Event e_1: A has appendicitis, B does not $Cr(e_1) = 0.16$	Event e_2: A does not have appendicitis, B does $Cr(e_2) = 0.16$	Event e_3: both A and B have appendicitis $Cr(e_3) = 0.64$	Event e_4: neither A nor B have appendicitis $Cr(e_4) = 0.04$
Give the medicine to A	A: 10 B: 10	A: 10 B: 1	A: 10 B: 1	A: 10 B: 10
Give the medicine to B	A: 1 B: 10	A: 10 B: 10	A: 1 B: 10	A: 10 B: 10
Split the medicine between A and B	A: 2 B: 10	A: 10 B: 2	A: 2 B: 2	A: 10 B: 10

[10] To say that these claims are probabilistically independent means that your credence that A has appendicitis is the same as your conditional credence that A has appendicitis given that B has appendicitis. Learning that B has appendicitis would not increase your credence that A has appendicitis: you don't think, for example, that appendicitis is catching so that one person having it is evidence that other people nearby have it too. And learning that B has appendicitis would not decrease your credence that A has appendicitis: you don't think, for example, that there is only so much appendicitis to go around. Learning that B has appendicitis would neither increase nor decrease your credence that A has appendicitis. And from this claim the reverse automatically follows: learning that A has appendicitis would not change your credence that B has appendicitis. The proposition that A has appendicitis and the proposition that B has appendicitis are probabilistically independent in this sense.

[11] There are interesting questions over what exactly we should be measuring here. The individuals' preferences? The preferences that they would have if fully informed? Or some other measure of their welfare? I set this question aside here and just assume that we are able to state the welfare of any outcome for any individual using a number on an interval scale.

WELFARE ECONOMICS 127

Obviously here we cannot simply apply the rule MEU, because each outcome is not assigned a single value, but is rather assigned a value for each member of the population affected. How, then, ought you to act in cases like this, where the choice of action affects the welfare of a group or population? What factors ought to be considered? These are the sorts of questions that are addressed by welfare economists, social choice theorists, and those working in distributive justice. And the answers given have enormous practical implications for all those involved in policy choice. There are a broad range of proposed rules for selecting between the options, but here I will just give a rough-and-ready outline of a few of these rules.[12]

6.5.1 Utilitarianism, egalitarianism, and prioritarianism

I begin with a version of utilitarianism: you ought to choose whichever action (or from amongst the actions, if there is a tie) that results in the highest expected total welfare—where the total welfare is simply the sum of the welfare for each member of the population. Thus in our example, at event e_1 the total welfare produced by giving the medicine to A would be $(10 + 10) = 20$. And the expected total welfare produced by giving the medicine to A is $(0.16)(20) + (0.16)(11) + (0.64)(11) + 0.04(20) = 3.2 + 1.76 + 7.04 + 0.8 = 12.8$. The expected total welfare produced by giving the medicine to B is similarly 12.8. And the expected total welfare produced by splitting the medicine between A and B is 7.2. Thus there are two actions that have the highest expected total welfare—giving the medicine to A and giving the medicine to B—and according to this version of utilitarianism you must choose one of those two actions.[13]

---

[12] See (Adler, 2017) for a clear and thorough overview.

[13] The calculation of *expected* welfare depends on the credence assigned to each state by the decision maker. This might give us pause: surely decisions of this nature ought to be made on the basis of objective probabilities rather than on the basis of an individual's credences? The problem with this idea is that typically the decision maker will not know the objective probabilities of each event, so cannot choose on the basis of these sorts of probabilities. Where the decision maker does know the objective probabilities, then plausibly those are the probabilities that she should work with—but in such cases the decision maker's credences rationally ought to match the objective probabilities in any case (more or less—some details relating to the Principal Principle are discussed in chapter 5). Where the decision maker does not know the objective probabilities, then plausibly she should work with her estimation—or to be more precise, her expectation—of the objective probabilities. And the Principal Principle (again, more or less) ensures that a rational agent's credences match up with her expectations of the objective probabilities.

128 PRACTICAL IMPLICATIONS

One sort of theorist who rejects this version of utilitarianism is the egalitarian. For the egalitarian, it is not just the total welfare that matters, but also how that welfare is arranged—and in particular how evenly or fairly it is arranged (Fleurbaey, 2008). A certain sort of egalitarian might see splitting the medicine as better than giving the medicine to either A or B, on the grounds that splitting the medicine is likely to lead to a more equal distribution of welfare, and this equality may compensate for the drop in overall welfare: at event e_3, for example, under the action of giving the medicine to A, there is quite a lot of total welfare (11) but it is very unequally distributed, whereas under the action of splitting the medicine there is less total welfare (4) but it is perfectly equally distributed. Egalitarians vary over how equality is to be weighed against other considerations, and in the details of their accounts. An alternative view—prioritarianism—holds that the welfare of those who are worse off should be prioritized over those who are better off (Parfit, 1991). Thus a prioritarian might also have a reason to see splitting the medicine as better than giving the medicine to either A or B: the outcome at e_3, for example, may be judged better under the action of splitting the medicine than under the other two actions, for though this choice reduces the welfare for the best off, this reduction may be more than compensated by the gain in welfare for the worst off. Again, different prioritarians spell out their account in different ways. In short, there are a wide variety of different rules for selecting between these actions, and this quick overview has just given a taste of a few.

6.5.2 The Pareto principle

Underlying the debate over these different rules are several principles, and these principles are sometimes used to argue for, define, or reject a particular sort of rule. One important such principle is the Pareto principle, which has two versions—one concerning outcomes and one concerning actions. I begin with the principle as it concerns outcomes. Take two outcomes, O_1 and O_2, with welfare assigned to the same population at each. Which outcome, O_1 or O_2, is the best? Well, let's suppose that for each person the welfare assigned to that person at O_1 is at least as great as the welfare assigned at O_2; and suppose in addition that for at least one person the welfare assigned to that person at O_1 is greater than the welfare assigned at O_2. In that case, outcome O_1 is said to be Pareto superior to outcome O_2, and the relevant Pareto principle states that outcome O_1 is therefore better than outcome O_2. To illustrate this principle, suppose that at one outcome two people are each assigned 1 unit of welfare, but at another outcome the first person is assigned 1 unit of welfare and the second 100 units of welfare. The second outcome is Pareto superior to the first—and so (according to this Pareto principle)—*better*. This principle may run into conflict with a certain sort of egalitarian, who might count the first outcome as better than the second on the grounds that it is more equal.

The second Pareto principle that we will consider is called the *ex ante* Pareto principle. This principle concerns the expected welfare (the 'prospects') for a person under a given action. To see what is meant by this, consider again our example of the appendicitis patients, this time with a calculation of each person's prospects included as a final column in the table (Table 6.15). A person's prospects under an action are given by the expected welfare of that person under that action. We can take as an example the prospects for A under the action of giving the medicine to A, which are calculated as follows: $(0.16)(10) + (0.16)(10) + (0.64)(10) + (0.04)(10) = 10$.

I can now explain what it is for one action to be *ex ante* Pareto superior to another: an action A_1 is *ex ante* Pareto superior to an action A_2 iff for every person the prospects under A_1 are at least as good as the prospects under A_2, and in addition, for at least one person, the prospects under A_1 are better than the prospects under A_2. In short, by switching to an action that is *ex ante* Pareto superior, we harm no one's prospects, and improve at least one person's prospects. And the *ex ante* Pareto principle states that it is not permissible to perform an action when some other action that is *ex ante* Pareto superior is available. We can apply this principle to our example involving the appendicitis patients in Table 6.15. We can see at once that no action is Pareto superior to the action of giving the medicine to A— for switching to any other action would harm A's prospects: thus as far as the *ex ante* Pareto principle rules, it is permissible to give the medicine to A. Similarly, it is permissible to give the medicine to B, for switching to either of the other actions would harm B's prospects; and it is permissible to split the medicine, for switching to either of the other actions would harm either A's or B's prospects. Thus the *ex ante* Pareto principle in this case leaves all actions permissible.

Many theorists find the *ex ante* Pareto principle compelling, and it plays a central role in the literature in welfare economics and social choice theory. Most famously, it has been used to construct arguments for utilitarianism (Harsanyi, 1977) and against any form of social evaluation that gives additional weight to the interests of those who end up worse off.

Table 6.15

	Event e_1: A has appendicitis, B does not $Cr(e_1) = 0.16$	Event e_2: A does not have appendicitis, B does $Cr(e_2) = 0.16$	Event e_3: both A and B have appendicitis $Cr(e_3) = 0.64$	Event e_4: neither A nor B have appendicitis $Cr(e_4) = 0.04$	Prospects
Give the medicine to A	A: 10 B: 10	A: 10 B: 1	A: 10 B: 1	A: 10 B: 10	A: 10 B: 2.8
Give the medicine to B	A: 1 B: 10	A: 10 B: 10	A: 1 B: 10	A: 10 B: 10	A: 2.8 B: 10
Split the medicine between A and B	A: 2 B: 10	A: 10 B: 2	A: 2 B: 2	A: 10 B: 10	A: 3.6 B: 3.6

WELFARE ECONOMICS 131

Those who are concerned with defending views that incorporate special concern for the worse off have therefore attacked this principle in its unrestricted form but, in a nod to its persuasive power, have attempted to accommodate it in a restricted form, by limiting it to cases in which there are no inequalities in final well-being (Fleurbaey, 2010; Fleurbaey and Voorhoeve, 2013). Philosophers have appealed to restricted versions of the *ex ante* Pareto principle in order to adjudicate debates among egalitarians and prioritarians. Egalitarians have argued that only they can respect the restricted form of the *ex ante* Pareto principle that applies just to situations without inequality (Voorhoeve and Otsuka, 2018)—and see (Adler and Holtug, 2019) for a reply. The principle thus plays a prominent role in these debates.

The *ex ante* Pareto principle of course rests on the idea of individuals having *prospects*—as indeed do several other important principles and positions in the literature on welfare economics. I argue that the very idea of prospects for an individual is wrong-headed, and ultimately this is because credence claims are opaque. In the next section I explain why.

6.5.3 Prospects for individuals and opacity

I have argued that credence claims are opaque. So it may be that your credence that a is F is different from your credence that b is F—even if a and b refer to the same object, and even if they both refer rigidly. From here it follows that where a and b are two names for the same person, a and b can have different prospects under the same action.

To see why this is, let's return to our example of the two appendicitis patients, whom you know as Alice and Belinda. Let us suppose that you hear another doctor discussing them, but she refers to them as 'Ms Smith' and 'Ms Jones'. You are sure that either Ms Smith is Alice and Ms Jones is Belinda, or vice versa, but you don't know which, and your credence is equally divided between these two possibilities. Of course the true identity claims are metaphysically necessary (e.g. if Alice is Ms Smith then this holds necessarily) but nevertheless you do not know which identity claims hold and have a credence of 0.5 in each possibility. This is exactly the sort of uncertainty that we should expect given that credence claims are opaque. Let us then add this extra uncertainty to our decision table (Table 6.16), writing 'AM' to mean that Alice has appendicitis, '¬AM' to mean that Alice doesn't have appendicitis (and so on), and 'A = S' to mean that Alice is Ms Smith

Table 6.16

	e_1: AM ¬BM A=S $Cr(e_1)$ = 0.08	$e_1^\star$: AM ¬BM A=J $Cr(e_{1^\star})$ = 0.08	e_2: ¬AM BM A=S $Cr(e_2)$ = 0.08	$e_2^\star$: ¬AM BM A=J $Cr(e_{2^\star})$ = 0.08	e_3: AM BM A=S $Cr(e_3)$ = 0.32	$e_3^\star$: AM BM A=J $Cr(e_{3^\star})$ = 0.32	e_4: ¬AM ¬BM A=S $Cr(e_4)$ = 0.02	$e_4^\star$: ¬AM ¬BM A=J $Cr(e_{4^\star})$ = 0.02	Prospects
Give the medicine to A	A: 10 B: 10 S: 10 J: 10	A: 10 B: 10 S: 10 J: 10	A: 10 B: 1 S: 10 J: 1	A: 10 B: 1 S: 1 J: 10	A: 10 B: 1 S: 10 J: 1	A: 10 B: 1 S: 1 J: 1	A: 10 B: 10 S: 10 J: 10	A: 10 B: 10 S: 10 J: 10	A: 10 B: 2.8 S: 6.4 J: 6.4
Give the medicine to B	A: 1 B: 10 S: 1 J: 10	A: 1 B: 10 S: 10 J: 1	A: 10 B: 10 S: 10 J: 10	A: 10 B: 10 S: 10 J: 10	A: 1 B: 10 S: 1 J: 10	A: 1 B: 10 S: 10 J: 1	A: 10 B: 10 S: 10 J: 10	A: 10 B: 10 S: 10 J: 10	A: 2.8 B: 10 S: 6.4 J: 6.4
Split the medicine between A and B	A: 2 B: 10 S: 2 J: 10	A: 2 B: 10 S: 10 J: 2	A: 10 B: 2 S: 10 J: 2	A: 10 B: 2 S: 2 J: 10	A: 2 B: 2 S: 2 J: 2	A: 2 B: 2 S: 2 J: 2	A: 10 B: 10 S: 10 J: 10	A: 10 B: 10 S: 10 J: 10	A: 3.6 B: 3.6 S: 3.6 J: 3.6

WELFARE ECONOMICS   133

and Belinda Ms Jones, and 'A = J' to mean that Alice is Ms Jones and Belinda Ms Smith.

To understand this table, let's focus as an example on event e_2, and the action of giving the medicine to Alice.[14] At event e_2, Alice does not have appendicitis but Belinda does, and Alice is Ms Smith and Belinda is Ms Jones. Under the action of giving the medicine to Alice, we can read off the outcome: Alice has high welfare (10) because Alice does not have appendicitis; Belinda has low welfare (1) because Belinda has appendicitis and does not get the medicine; Ms Smith has high welfare (10), because Ms Smith is Alice and so does not have appendicitis; and Ms Jones has low welfare (1), because Ms Jones is Belinda and so has appendicitis and does not get the medicine.

In the final column of the table, we can see the prospects for Alice and Belinda, and for Ms Smith and Ms Jones under each action. The first thing to note is that under a single action, the prospects for an individual can be different depending on how that individual is designated. Let us suppose, for example, that though you do not know it, in fact Alice is Ms Smith—so 'Alice' and 'Ms Smith' designate a single individual. Nevertheless, the prospects for Alice (10) under the action of giving the medicine to A are different from the prospects for Ms Smith (6.4) under that very same action. Thus it seems that there is no such thing as the prospects for an individual under an action; rather, we have prospects under an action for an individual *designated in a particular way*. For example, there are prospects for Ms Smith (so designated) and prospects for Alice (so designated), and these may come apart even if in fact Ms Smith is Alice.

This creates problems for the many principles and arguments in the literature that rely on the idea of prospects for individuals. Here I will focus on showing how it is a problem for the *ex ante* Pareto principle. Let's think again about our example of the two appendicitis patients. Earlier we were focused on these two patients under their designators 'Alice' and 'Belinda', and it seemed that none of the three possible actions were *ex ante* Pareto superior to all the others: giving the medicine to Alice is not *ex ante* superior, because we could improve the prospects for Belinda by

---

[14] In a sense, by giving the medicine to Alice you will be giving the medicine to Ms Smith—if it happens to be the case that Alice is Ms Smith. But we should treat these as distinct actions (for reasons explained in section 6.3—where I argued that actions are opaque). In this example, we're assuming that the option of giving the medicine to Ms Smith is not open to you—presumably because you are unable to make this true at will—so this option does not appear in the decision table.

134 PRACTICAL IMPLICATIONS

switching to either of the other actions; and similarly giving the medicine to Belinda is not *ex ante* superior because the prospects for Alice can be improved by switching to either of the other actions; and splitting the medicine between the patients is also not *ex ante* Pareto superior, because the prospects for each of Alice and Belinda can be improved by switching to one of the other two actions. But now we have introduced an alternative pair of names for designating the two patients concerned: Ms Smith and Ms Jones: what happens if we focus on the prospects for the patients designated in these ways instead? Then the actions of giving the medicine to Alice and giving the medicine to Belinda emerge as *ex ante* Pareto superior to the action of splitting the medicine. Under the action of splitting the medicine, both Ms Smith and Ms Jones have prospects of 3.6, whereas under either of the other two actions they both have better prospects of 6.4. Splitting the medicine thus has worse prospects for both Ms Smith and Ms Jones than their prospects under either of the other two actions, and so by the *ex ante* Pareto principle the action of splitting the medicine is impermissible, and one of the other two actions must be chosen.

This is a strange and troubling result. It seems that whether one action is *ex ante* Pareto superior to another depends on how the people concerned are designated: an action might be *ex ante* Pareto superior relative to one set of designators, but not relative to another. From here we have a number of options. A natural first thought is that we should have a rule for identifying the right set of designators. But what should that rule be? We might try limiting the designators to those that are rigid, but as we can see from the example there can be two or more sets consisting just of proper names, which are rigid designators. We might focus on those designators that the people concerned would recognize as their own—but again in our example there are two such sets (for example, if Alice is Ms Smith, then we can assume that she knows both that she is Alice and that she is Ms Smith). We might allow intuition or moral considerations to guide our choice of the relevant designators, but this runs counter to the spirit of the *ex ante* Pareto principle: to many, the principle seems obviously right (why choose action A when switching to action B hurts nobody's prospects, and improves the prospects for somebody?) but any application of the principle will be up for debate if the set of relevant designators has first to be fixed by intuition or moral considerations.

This problem also arises if we attempt to mandate an official designator. Suppose, for example, that it is mandated that it is prospects for individuals *under passport numbers* that should be considered. One practical problem with this idea is that there are people who lack passport numbers (babies, for example) and so will be excluded from consideration altogether, and there are also people who have two such numbers. But there is a further problem, which is that even when an individual does have a unique passport number, the relevant decision maker may not know what it is: the decision maker may have highly relevant information about a person without being able to connect this to any particular passport number. As an example, take a doctor attending to emergency patients at a hospital. A patient arrives, unconscious, and with no identifying information. The doctor can see at once that he can vastly improve the prospects *for this patient* by giving a particular treatment. But—as the doctor has no idea of the patient's passport number—there is no person identified by passport number for whom the treatment would make any more than negligible difference to their prospects.

In short, there is no easy way to identify an appropriate set of designators here.[15]

6.5.4 Supervaluationism

My preferred response to this problem is to take a supervaluationist approach. I explore the implications of this in depth elsewhere (Mahtani, 2017; Mahtani, 2021), and below I give just a brief introduction to the approach.

We know what it means for one action to be *ex ante* Pareto superior to another *relative to a set of designators*. On the supervaluationist approach, we then say that one action is *ex ante* Pareto superior to another *simpliciter*, iff it is *ex ante* Pareto superior relative to *every* set of designators: every set of designators, that is, such that the decision maker is certain that it contains one and only one designator for each of the people concerned. Thus, for example, in our earlier case of the two appendicitis patients, no action emerges as *ex ante* Pareto superior *simpliciter*: splitting the medicine

---

[15] Here you might wonder whether we could say quite generally that in welfare economics recommendations for action should always be read as *relative to a set of designators*. But then we would be left with the question of how to act. It might be that one action is recommended relative to one set of designators, and another action is recommended relative to another set of designators. You cannot act in one way (say, give the medicine to A) relative to one set of designators, and also act in a different way (say, split the medicine) relative to another set of designators. You don't act relative to a set of designators—you simply act! See footnote 14 in chapter 4 for a related point about action relative to guises.

136 PRACTICAL IMPLICATIONS

is *ex ante* Pareto superior relative to the set {Ms Smith, Ms Jones}, but it is not *ex ante* Pareto superior *simpliciter*, for that would require it to be *ex ante* Pareto superior relative to every set of designators, and it is not *ex ante* Pareto superior relative to the set {Alice, Belinda}.

Taking the supervaluationist approach to the *ex ante* Pareto principle makes some important changes to the landscape of welfare economics. Previously we might have taken one action to be *ex ante* superior to another just because it was *ex ante* Pareto superior relative to the most obvious set of designators—but now it will count as *ex ante* Pareto superior only if it is so relative to *every* set of designators. The result is that it is now harder for one action to qualify as *ex ante* Pareto superior to another, and so the *ex ante* Pareto principle is weaker, for it rules fewer actions out as impermissible. Now that the principle is seen to be weaker than previously thought, it becomes clear that it cannot play all the roles that had been expected of it. In particular, it cannot play the same role in arguments for utilitarianism, nor in adjudicating between versions of egalitarianism and prioritarianism.

Thus the fact that credence claims are opaque has some wide-reaching implications in welfare economics. The very idea of prospects for individuals needs to be rethought, as the same individual can have different prospects under different designators. As we have seen, this means that the *ex ante* Pareto principle needs to be amended, and this has some serious ramifications. And this is just an example: there are many other positions and arguments in welfare economics that make use of the idea of prospects for individuals, and many of these will need to be revisited. It seems possible, then, for welfare economics to accommodate the fact that credence claims are opaque, but the implications will be wide-reaching.

6.6 Chapter summary

I began by introducing decision theory, focusing on the popular decision rule 'maximize expected utility' (MEU). It seems that this framework can accommodate my tenet that credence claims are opaque, but once this tenet is explicitly considered we can see that there are a number of important implications. I looked at two such implications: one concerning the two-envelope paradox, and another relating to the concept of prospects in welfare economics. These examples illustrate the importance of the tenet that credence claims are opaque: if this is not borne firmly in mind, then we are apt to go astray in our decision-making and policy choice.

Having spent this part of the book arguing for the importance of the tenet that credence claims are opaque, I turn now to the final part, where I explore this foundational question: how can the Bayesian framework accommodate the claim that credence claims are opaque?

7

States as Metaphysically Possible Worlds

7.1 Introduction

So far in this book I've argued that credence claims are opaque, and uncovered some important and surprising implications of this claim. In this final part of the book, I turn to foundations: can the framework accommodate the fact that credence claims are opaque? How?

To see the issue here, let's recall the probability framework as described in chapter 3. A 'probability space' consists of a set of states (Ω), an algebra ($\mathcal{F}$) over that set of states, and a probability function on $\mathcal{F}$ (P). As explained in chapter 3, a probability space is an abstract, mathematical object, but it can be interpreted in various ways. Those who work with the credence framework interpret a probability space as an agent's epistemic state. On this interpretation, a probability function assigns values to the elements of the algebra, and these values represent the credences that the agent has in various propositions. On this interpretation, then, propositions correspond to elements of the algebra, and each element of the algebra is a set of states. Is this compatible with the fact that credence claims are opaque? That depends on how, in more detail, we spell out the interpretation: on a natural way of spelling out the details, we face a problem.

Here's the natural way of spelling out the details: we take states to be metaphysically possible worlds, and then each element of the algebra—and so each proposition—is a set of metaphysically possible worlds. This aligns well with the account of propositions given by theorists who endorse 'possible worlds semantics' (as discussed in chapter 2). But it seems to lead immediately to a problem. For consider that—given that credence claims are opaque—the following two sentences may both be true:

(7a)   Tom has a credence of 0.8 that George Orwell is a writer.

(7b)   Tom has a credence of 0.1 that Eric Blair is a writer.

These two sentences attribute certain credences to Tom. Sentence (7a) attributes to Tom a credence of 0.8 in the proposition expressed by

The Objects of Credence. Anna Mahtani, Oxford University Press. © Anna Mahtani 2024.
DOI: 10.1093/oso/9780198847892.003.0007

INTRODUCTION 139

'George Orwell is a writer', and sentence (7b) attributes to Tom a credence of 0.1 in the proposition expressed by 'Eric Blair is a writer'. On the view that we're considering, a proposition corresponds to a set of possible worlds. The proposition expressed by 'George Orwell is a writer' presumably corresponds to the set of possible worlds at which George Orwell is a writer; and the proposition expressed by 'Eric Blair is a writer' presumably corresponds to the set of possible worlds at which Eric Blair is a writer. But the set of possible worlds at which George Orwell is a writer is exactly the same as the set of possible worlds at which Eric Blair is a writer. Thus it seems that sentences (7a) and (7b) attribute different credences to Tom in the very same proposition. Given that on the credence framework Tom's epistemic attitude is represented by a credence *function*, it seems that (7a) and (7b) are inconsistent. Thus this natural way of spelling out the details of the credence framework is incompatible with the fact that credence claims are opaque.

There are broadly two ways that we might respond to this problem. One broad strategy is to drop the idea that states are metaphysically possible worlds. Perhaps a state is something more fine-grained, for example? This sort of option will be explored in the next chapter (chapter 8). The other broad strategy is to maintain that states are metaphysically possible worlds, and question some other move in the reasoning above. This is the strategy that I explore in this chapter. I begin (in 7.2) by questioning whether it is really the case that the set of metaphysically possible worlds at which George Orwell is a writer is the same as the set of metaphysically possible worlds at which Eric Blair is a writer. I consider a view based on Russell's descriptivism that rejects this claim, but that view faces numerous well-known problems.[1] I then try questioning a different assumption in the reasoning above. The reasoning above involved this claim: 'the proposition expressed [in 7a] by "George Orwell is a writer" presumably corresponds to the set of possible worlds at which George Orwell is a writer'. But why couldn't it correspond to a different set of possible worlds? The expression 'George Orwell is a writer' in (7a) is directing us to *some* proposition, which (on the view that we're exploring) must correspond to a set of

---

[1] There are various other views that would fall within this same broad approach—i.e. other views on which the set of metaphysically possible worlds at which George Orwell is a writer is not the same as the set of metaphysically possible worlds at which Eric Blair is a writer. One such view arises from counterpart theory (Lewis, 1973, pp. 39–43) according to which an object has counterparts at various possible worlds, rather than existing at each of them. Such a view may have the resources to handle our problem, but it comes with certain controversial implications (Mark Jago, personal correspondence). I do not attempt to explore this view here.

140   STATES AS METAPHYSICALLY POSSIBLE WORLDS

metaphysically possible worlds—but must it be the set of possible worlds where George Orwell is a writer? If not, then there is the hope that the expressions 'George Orwell is a writer' and 'Eric Blair is a writer' direct us to different sets of metaphysically possible worlds, and thus we can make sense of the claims that assign Tom different credences in each. I discuss two different but related versions of this approach: a version based on Robert Stalnaker's account of belief attribution (7.3), and a version based on Chalmers's two-dimensionalism (7.4–7.8). I spend a lot of this chapter on Chalmers's account, partly because the account is complex, and partly because it appears, at least at first sight, to be a promising solution to our problem. The discussions of Russell and Stalnaker are comparatively brief, and much briefer than these views deserve. But ultimately all the approaches that I consider face problems—not insurmountable problems perhaps—but problems that should certainly be borne in mind by users of the credence framework.

I begin in the next section with a brief reminder of Russell's descriptivism.

7.2 Russell's descriptivism

In chapter 2 I outlined Russell's theory of definite descriptions, and explained that on his view most ordinary proper names are definite descriptions in disguise. Russell himself held that propositions are *structured* entities—so not sets of possible worlds at all—but for the purposes of this section, I consider whether we can mix Russell's descriptivist account of proper names with a possible worlds account of propositions, in the hope that this will give us an account that can handle the fact that credence claims are opaque.

Recall the statements below, (7a) and (7b):

(7a)   Tom has a credence of 0.8 that George Orwell is a writer.

(7b)   Tom has a credence of 0.1 that Eric Blair is a writer.

If we take the objects of credence to be sets of metaphysically possible worlds, then it seems that (7a) and (7b) cannot both be true together—for the possible worlds at which George Orwell is a writer are exactly the same as the possible worlds at which Eric Blair is a writer. But why should we think that any possible world where George Orwell is a writer is also a world

RUSSELL'S DESCRIPTIVISM 141

where Eric Blair is a writer, and vice versa? Our reason for thinking this was that names are rigid designators: 'George Orwell' and 'Eric Blair' refer to the same person at every possible world. But on Russell's account, names are definite descriptions in disguise, and at least in some contexts definite descriptions do not designate rigidly.

Suppose, for example, that 'George Orwell' is equivalent to the description 'the winner of the 1946 Hugo Prize', and 'Eric Blair' is equivalent to the description 'the man born in 1903 who was christened "Eric Blair"'. In that case, we can write out (7a) and (7b) in their fuller forms as follows:

(7a')   Tom has a credence of 0.8 that the winner of the 1946 Hugo Prize is a writer.

(7b')   Tom has a credence of 0.1 that the man born in 1903 who was christened 'Eric Blair' is a writer.

On the view that we're considering, (7a') attributes to Tom a credence of 0.8 in the proposition that corresponds to the set of metaphysically possible worlds at which the 1946 Hugo Prize is a writer. And (7b') attributes to Tom a credence of 0.1 in the proposition that corresponds to the set of metaphysically possible worlds at which the man born in 1903 who was christened 'Eric Blair' is a writer. Given that these are two different sets of possible worlds, our puzzle seems to be resolved: Tom can obviously have different credences in two different propositions.

Of course, this approach inherits all the problems that have dogged Russell's descriptivist account of names (Kripke, 1980). If each proper name is a disguised definite description, then what description goes with each name? There is no obvious answer here. It might be that a particular description springs to your mind when you hear the name 'George Orwell'— but should we then say that this is *the* description that the name 'George Orwell' is equivalent to? What if other people associate different descriptions with the name? And what if there are names that you associate with no particular definite description at all? These are serious problems—though there are descriptivists who try to address them (Searle, 1982).

Perhaps the most influential problem raised against Russell's descriptivist account of names concerns rigidity (Kripke, 1980): if names are definite descriptions in disguise, then how come names are rigid designators and definite descriptions are not? That names are rigid designators seems undeniable (see section 2.6). The descriptivist can respond to this problem

142 STATES AS METAPHYSICALLY POSSIBLE WORLDS

by pointing out that definite descriptions have a rigid reading as well as a non-rigid reading, and, indeed, the rigid reading can be forced by adding an 'actually' to the description. Thus though 'the winner of the 1946 Hugo Prize' has a non-rigid reading, 'the *actual* winner of the 1946 Hugo Prize' has only a rigid reading. In general, the descriptivist can make her account more plausible by claiming that names are shorthand for rigidified (rather than non-rigidified) definite descriptions (see (Nelson, 2002, pp. 427, footnote 4) for an overview of this move in the literature). For our purposes, though, this move is disastrous. We hoped to use descriptivism to explain how Tom can have a high credence that George Orwell is a writer, and a low credence that Eric Blair is a writer. The thought was that if 'George Orwell' and 'Eric Blair' were both just descriptions in disguise, then the set of possible worlds at which George Orwell is a writer would be different from the set of possible worlds at which Eric Blair is a writer, and so we could make sense of Tom having a different credence in each. But if the descriptions have to be rigidified, then the sets of possible worlds will be identical. The set of possible worlds at which the *actual* winner of the 1946 Hugo Prize is a writer will be the set of worlds at which that very person (George Orwell) is a writer; and the set of possible worlds at which the person who was *actually* born in 1903 and christened 'Eric Blair' is a writer will be the set of possible worlds at which that very person (George Orwell) is a writer. This is the same set of possible worlds. By rigidifying the definite descriptions—to get them to behave more like proper names and so make it more plausible that they mean the same thing—we have lost our solution to the problem.

I set aside this attempt to use Russell's descriptivism to solve our puzzle. To pursue our strategy we will need to try a different approach, and I turn in the next section to Stalnaker's account of belief attribution.

7.3 Stalnaker's account of belief attribution

Stalnaker's views on representation range over a wide area, including philosophy of language, philosophy of mind, and epistemology. Here for our purposes, I pull out just one strand of his work, which focuses on belief attribution statements.

On Stalnaker's view, the object of a belief is a proposition, and a proposition corresponds to a set of metaphysically possible worlds. But this account seems to face a problem—the belief-analogue of the problem we are grappling with. Stalnaker draws out this problem by focusing on

STALNAKER'S ACCOUNT OF BELIEF ATTRIBUTION   143

belief attribution statements involving mathematical claims. Take the two mathematical claims 'two plus two is four' and 'there are an infinite number of primes'. Both of these mathematical claims are necessary: that is, they are both true at all metaphysically possible worlds. It seems to follow, then, that they express the same proposition. And yet, as Stalnaker points out, claims (7c) and (7d) below may both be true (Stalnaker, 1984, p. 73):

(7c)   O'Leary believes that two plus two is four.

(7d)   O'Leary does not believe that there are an infinite number of primes.

Part of Stalnaker's response to this problem is to interrogate (7d). Is (7d) really saying that O'Leary is ignorant of the necessary proposition? Not according to Stalnaker. Rather (7d) is saying that O'Leary does not know that 'there are an infinite number of primes' expresses the necessary proposition. Thus O'Leary's ignorance is really ignorance about a linguistic matter: he doesn't know what 'there are an infinite number of primes' expresses.

To grasp Stalnaker's move, we need to recognize a subtle distinction that Stalnaker draws between the 'semantic content' and the 'diagonal content' of an assertion. We can start by imagining an utterance of a particular sentence taking place (with assertoric force) in our actual world. For example, imagine that I utter the sentence 'grass is green'. The context of utterance helps to determine which proposition is expressed. Which proposition I have expressed might depend partly on the physical environment, partly on my intentions in making the utterance, and partly on the conventions of my linguistic community (that is, how the people around me usually use these words)—and this is all part of the context of utterance. With the proposition expressed thus settled by the actual-world context of utterance, we can then consider at which possible worlds that proposition is true. It will be true at worlds where grass is green, but false at worlds where grass is, say, blue. Here, then, the actual world is providing the context of utterance—fixing which proposition is expressed—and then various possible worlds are providing the contexts of evaluation as we consider the truth value of the proposition at each possible world. We then define the semantic content of an utterance as the set of possible worlds at which the proposition expressed by the utterance, given the *actual* context of utterance, is true.

Now let's change tack and imagine that the utterance of that sentence takes place at a range of different possible worlds. So, for example, we can

144 STATES AS METAPHYSICALLY POSSIBLE WORLDS

suppose that I in the actual world say 'grass is green', and at other possible worlds I also say 'grass is green'. I may not be expressing the same proposition at every possible world. For there are possible worlds where 'green' is used differently by the linguistic community, and so means blue rather than green, for example. At those possible worlds, an utterance of 'grass is green' might express the proposition that grass is blue. With this in mind, we can then understand the idea of the diagonal content (as opposed to the semantic content) of the utterance as follows.[2] At the actual world, I make an utterance of 'grass is green'. The context of utterance (the actual world) fixes which proposition is expressed—in this case the proposition that grass is green. Is that proposition true at the actual world? That is, is grass green at the actual world? It is, so the actual world belongs in the diagonal content. Now let's take another possible world. At that world I utter 'grass is green', and the context of utterance at that world fixes which proposition is expressed, which let's suppose is the proposition that grass is blue. Is that proposition true at that world? That is, is grass blue at the world where the utterance is made? If so (and only if so), then this world also belongs in the diagonal content. And so on for all possible worlds. This is how the diagonal content of the utterance is constructed: it contains all and only those possible worlds such that the utterance as made at that world expresses a proposition that is true at that world. Here we are treating each possible world as giving both the context of utterance and the context of evaluation.

The semantic content of an utterance may be different from the diagonal content of an utterance. For example, an utterance (at the actual world) of 'there are an infinite number of primes' has as its semantic content the set of all metaphysically possible worlds. In contrast, an utterance of 'there are an infinite number of primes' has as its diagonal content a set smaller than the set of all metaphysically possible worlds. For there are metaphysically possible worlds where, for example, 'infinite' means finite, and an utterance *at that world* of 'there are an infinite number of primes' expresses a proposition which is false at that world (and indeed at all worlds). This world, then, will not belong in the diagonal content, and so we can see that the diagonal content of the utterance is different from its semantic content: the semantic content includes all worlds, while the diagonal content contains fewer worlds.

---

[2] Stalnaker uses a matrix to explain his point, which is why he chooses the term 'diagonal content' here.

STALNAKER'S ACCOUNT OF BELIEF ATTRIBUTION   145

This subtle distinction helps resolve the problem with claim (7d). It looks like (7d) states that O'Leary does not believe the *semantic* content of 'there are an infinite number of primes', and this would be problematic. For the semantic content of 'there are an infinite number of primes' is the set of all possible worlds—the same as the semantic content of 'two plus two is four', which O'Leary does believe as (7c) seems to state. But there is an alternative reading of (7c) and (7d). We can take (7c) to be stating that O'Leary believes the diagonal content of 'two plus two is four', while (7d) states that O'Leary does not believe the diagonal content of 'there are an infinite number of primes'. As the diagonal content of 'two plus two is four' is different from the diagonal content of 'there are an infinite number of primes', this seems to resolve the puzzle. The crucial move is to recognize that in a belief attribution statement like (7d), the content of the belief is not—as might be assumed—the set of metaphysically possible worlds at which there are an infinite number of primes. For O'Leary's uncertainty is not really over whether there are an infinite number of primes, but rather over what content is expressed by 'there are an infinite number of primes': he doesn't know whether he is in a world where this sentence expresses the necessarily true proposition, or the necessarily false proposition.

This is part of Stalnaker's response to the problem of belief attribution statements when they involve mathematics. Stalnaker applies it cautiously: he states that 'the suggestion that the contents of mathematical attitudes are propositions about the relation between expressions and the one necessary proposition is *prima facie* more plausible in some cases than in others' (Stalnaker, 1984, p. 74). There is more to Stalnaker's understanding of mathematical belief than I have described here, and we are not given unrestricted licence to apply this strategy to other kinds of necessary truths and falsehoods (Stalnaker, 1984, p. 75). Nevertheless, even if it goes beyond Stalnaker's intentions, we can try using his account to address the central issue of this book—namely that credence claims are opaque.

Recall the two statements below, (7a) and (7b):

(7a)   Tom has a credence of 0.8 that George Orwell is a writer.
(7b)   Tom has a credence of 0.1 that Eric Blair is a writer.

On a natural interpretation of the credence framework, the objects of credence are sets of metaphysically possible worlds. And on a natural interpretation of credence attribution statements like (7a) and (7b), the

146 STATES AS METAPHYSICALLY POSSIBLE WORLDS

relevant objects of credence are the semantic contents of 'George Orwell is a writer' and 'Eric Blair is a writer' respectively. The trouble, as we have seen, is that these semantic contents are the same. Why not, then, say that the relevant objects of credence are the diagonal contents of these claims? Then the objects of credence will be different—for there will be possible worlds where 'George Orwell is a writer' *as uttered at that world* is true, but 'Eric Blair is a writer' *as uttered at that world* is false, and vice versa—and so the problem will be resolved.

Here we can raise a natural objection to the proposal (Williamson, 2011; Jago, 2014, pp. 59–64). The objection is that according to the proposal, the object of Tom's credence is at least partly linguistic. For example, on this proposal (7b) states that Tom has a credence of 0.1 in the set of possible worlds at which words are used in such a way that the sentence 'Eric Blair is a writer' comes out as true. Thus Tom's credence concerns not just matters of fact, but also matters of language. But intuitively Tom's credence is not about such linguistic matters at all. To see this, let's suppose that Tom has been consciously wondering whether Eric Blair is a writer, and as a result of this process of wondering he arrives at his 0.1 credence. What sorts of considerations led to this credence? Let's suppose—as before—that Tom works in a café where Eric Blair frequently visits. Tom might have considered how his customer behaves when he's in the café—how he speaks and whether he ever writes anything down, or looks particularly thoughtful. Plausibly, Tom is not wondering about linguistic matters. He is not wondering who the name 'Eric Blair' refers to (he thinks he knows the answer to that), and he is not wondering what the word 'author' means (again, he thinks he knows that). It just seems implausible, then, to claim that the object of Tom's credence is even partly about linguistic conventions.[3]

Here, then, I set Stalnaker's account to one side, and turn to an alternative—Chalmers's two-dimensionalist account.

7.4 Chalmers's account—the basics

Chalmers has directly addressed a question that is central to this book: what are the objects of credence? In this chapter I describe and develop Chalmers's answer in his key paper on the topic (Chalmers, 2011a) and also

---

[3] Plausibly, Tom knows that 'Eric Blair is a writer' means that Eric Blair is a writer. Arguably this point in itself undermines Stalnaker's strategy (Field, 1978, p. 15; Speaks, 2006, pp. 448–50).

CHALMERS'S ACCOUNT—THE BASICS   147

draw on some variations and extensions of his position that can be found elsewhere.[4] I argue that there is no easy way to integrate Chalmers's account into the credence framework. On any plausible version of Chalmers's account, we can incorporate it only by a radical transformation.

I begin by giving a rough overview of Chalmers's two-dimensionalism account, starting with two different ways of thinking of possible worlds: we can consider them *as counterfactual*, or we can consider them *as actual* (Jackson, 2004). To philosophers, the familiar way of thinking about possible worlds is as counterfactual. When we consider possible worlds as counterfactual, then the actual world keeps its special status, and *actually P* obtains (at any given possible world) iff P holds at the actual world. So if P obtains at the actual world, but there is some non-actual world w where P does not obtain, then at w, P does not obtain, but *actually P* does. But things look rather different if we consider world w as actual, rather than as counterfactual. To consider w as actual is to consider the hypothesis that we are actually at w—that our description of w is a description of how things really are. Under this hypothesis, P is false, and *actually P* is false too. Thus there are two different ways of thinking about possible worlds, and what we would say obtains at a given world can depend on whether we are considering it as counterfactual or as actual.

Corresponding to each assertion, then, are *two* sets of possible worlds. First, there is the set of worlds (considered as counterfactual) at which the assertion is true. This is a familiar idea—it is the standard account of content according to possible worlds semantics—and Chalmers calls this set of worlds the 'secondary intension' of the assertion. But corresponding to an assertion there is also another set of possible worlds. These are the worlds where the assertion obtains *if those worlds are considered as actual*. As Chalmers puts it, these are the worlds that 'verify' the assertion. World w verifies an assertion A iff under the hypothesis that world w is actual, A follows a priori. For example, consider a world w where grass is blue and the sky is green. Under the hypothesis that world w is actual, it follows a priori that grass is blue, and so world w verifies the assertion that grass is blue. Furthermore, under the hypothesis that world w is actual, it follows a priori that *actually* grass is blue, so this assertion is also verified. The set of

---

[4] In (Chalmers, 2011a) Chalmers does not just put forward his own views about the objects of credence, but also argues that the objects must be non-referential, argues further that this is a reason to think that propositions quite generally must be non-referential, and criticizes various alternatives to his own account of the objects of credence. Here I focus on the parts of the paper where Chalmers puts forward his own views on the objects of credence.

148 STATES AS METAPHYSICALLY POSSIBLE WORLDS

worlds that verify an assertion is called the 'primary intension' of the assertion. And we can see that the primary intension of an assertion can be different from its secondary intension. For example, the assertion that actually grass is blue is true at no worlds (considered as counterfactual), and so its secondary intension is the empty set; but the assertion is nevertheless verified at some worlds, and so it has a non-empty primary intension.

On Chalmers's view, then, each assertion is associated with two sets of metaphysically possible worlds—its primary and secondary intensions—and this is why the view is called 'two-dimensionalism'. There is a complication here—for Chalmers claims that the sorts of worlds that belong in an assertion's primary intension are not quite the same as the sorts of worlds that belong in an assertion's secondary intension. The worlds that belong in an assertion's secondary intension are ordinary uncentred metaphysically possible worlds, but the worlds that belong in an assertion's primary intension are 'epistemically possible scenarios'. Epistemically possible scenarios are roughly equivalent to centred possible worlds—where a centred possible world (as described in chapter 2) is an ordinary metaphysically possible world plus a centre consisting of a person and a time. This is no doubt a simplification, and some thoughts on what epistemically possible scenarios are can be found in (Chalmers, 2011b), but for our purposes in this chapter we can treat epistemically possible scenarios simply as centred possible worlds.

Let us now see how Chalmers's two-dimensionalist approach applies to our familiar example of George Orwell and Eric Blair. George Orwell is Eric Blair, and so this identity holds at every metaphysically possible world where this person exists. For example, consider a world w^* at which Eric Blair never becomes a writer at all, and never takes on any pen name, while some other man calls himself 'George Orwell' and writes *Animal Farm*, *1984*, and so on. Even at world w^* (considered as counterfactual) it is true that George Orwell is Eric Blair: this is just a world where George Orwell/Eric Blair did not call himself 'George Orwell', or write any books. So it is true at w^* that George Orwell is Eric Blair. But is that assertion *verified* at w^*? To answer this question, we need to consider the hypothesis that w^* is actual—that things really are as the description of w^* suggests. Let us suppose, then, that it really is the case that Eric Blair never became a writer at all, and it is some other man who took on the pen name of 'George Orwell' and wrote *Animal Farm*, *1984*, and so on. Surely, under the supposition that that is how things actually are, it turns out that George Orwell is *not* Eric Blair after all. On Chalmers's view, this follows a priori: we associate the name

CHALMERS'S ACCOUNT—THE BASICS 149

'George Orwell' with certain a priori connections (taking on the pen name 'George Orwell', writing *Animal Farm* and *1984* and so on),[5] and in w^* as described these a priori connections attach to a person other than Eric Blair. From the supposition that w^* is actual, it follows that George Orwell is not Eric Blair, and thus w^* verifies the assertion that George Orwell is not Eric Blair. We can see, then, that an assertion of 'George Orwell is not Eric Blair' has a different primary and secondary intension. Its secondary intension is the empty set, but its primary intension is non empty, because there are worlds where the assertion is verified. Furthermore, the assertions 'George Orwell is a writer' and 'Eric Blair is a writer' have different primary intensions: world w^* verifies the assertion that George Orwell is a writer (and so belongs in its primary intension), but does not verify the assertion that Eric Blair is a writer (and so does not belong in its primary intension). This suggests some promising ways in which the credence framework might be able to accommodate the central claim of this book—namely that credence claims are opaque. I turn to these in the next section, but first I end this section with two important clarifications of the two-dimensional approach.

First, a clarification about the idea of verification. Recall that a world w verifies an assertion A iff under the hypothesis that world w is actual, A follows a priori. When considering the hypothesis that you inhabit some world, it matters how that world is described. And there are various ways that we might describe one and the same world. For example, in the last paragraph I described w^* and stated that under the hypothesis that w^* is actual, it follows a priori that George Orwell is not Eric Blair. But now suppose that we describe w^* further by adding that it is a world in which George Orwell and Eric Blair are one and the same. This is a perfectly accurate addition to the description, for it is indeed true that at w^* (and at every world where he exists) George Orwell is Eric Blair. But under the hypothesis that you inhabit w^*—described as before but with the additional detail that George Orwell is Eric Blair—it is no longer clear whether it follows a priori that George Orwell is not Eric Blair. Thus it seems that w^* does not straightforwardly verify the assertion after all. To deal with this sort of problem Chalmers introduces the idea of a 'canonical specification' of a world, which is (roughly) a complete description of the world given only in neutral vocabulary, where neutral vocabulary excludes proper names such as

---

[5] This description of the a priori connections are over-simplified in many ways, but they should help explain the rough idea.

150 STATES AS METAPHYSICALLY POSSIBLE WORLDS

'George Orwell' and 'Eric Blair'. Thus our above addition to the description of w^* is not allowed. The definition, then, is that a world w verifies an assertion A iff from the hypothesis that the world w—described using its canonical specification—is actual, A follows a priori.

The second clarification is to note that the primary and secondary intension of an assertion depends on the context in which the assertion is made. A sentence *type* is not the sort of thing that (usually) has a primary and secondary intension. Rather, it is an assertion—a particular utterance (with assertoric force) of a sentence in a certain context—that has a primary and secondary intension. To illustrate this, first let's start by focusing on secondary intensions, and consider the sentence 'I am hungry'. If Alice utters this sentence then it will have one secondary intension (the set of possible worlds in which Alice is hungry), and if Bob utters this sentence it will have a different secondary intension (the set of possible worlds in which Bob is hungry).[6] It is generally accepted that the proposition expressed by an utterance depends not just on the sentence type, but on the context—e.g. who uttered the sentence, at what time, in what environment, and so on—so it is not surprising to find that the secondary intension of an assertion depends on the context of utterance. It may be harder to see that primary intensions also depend on context, but they do. For example, if I say 'Jessie is tired', meaning Jessie my mother, then the primary intension of my assertion will be the set of centred possible worlds where the person I grew up with, who looks a certain way and so on, is tired. When you say 'Jessie is tired', meaning Jessie J the singer, then the primary intension of your assertion will be a different set of centred possible worlds—the set of centred possible worlds where the singer of 'Price Tag', and so on, is tired. We are each using the name 'Jessie' with different a priori connections, and of course in this example our connections lead to two different people. But in some cases two people can use the same name with different a priori connections even though both uses of the name refer to the very same individual. Here is an example:

> [I]t is often the case that two speakers will use the same *name* with different *a priori* connections. The canonical case is that of Leverrier's use of 'Neptune', which he introduced as a name for (roughly) whatever perturbed the orbit of Uranus. For Leverrier, 'If Neptune exists, it perturbs the

---

[6] Recall that the secondary intension of an assertion is the set of possible worlds (not centred possible worlds) at which the assertion is true.

orbit of Uranus' was *a priori*. On the other hand, later speakers used the term (and still do) so that this sentence is not *a priori* for them: it is epistemically possible for me that Neptune does not perturb the orbit of Uranus. We can even imagine that when Leverrier's wife acquired the name, she did not acquire the association with Uranus, so that she is in no position to know the truth of this sentence *a priori*.

(Chalmers, 2004, p. 203)

Chalmers explains that Leverrier and his wife can both utter the same sentence—say 'Neptune is an asteroid'—and their assertions will have different primary (in this context, Chalmers calls them 'epistemic') intensions: the scenarios that verify their assertions differ. This sort of variation in the primary intension can occur whenever a sentence contains proper names, natural kind terms (such as 'water' and 'gold'), demonstratives (such as 'that' and 'there'), and various other context-dependent terms (Chalmers, 2004, p. 207). To summarize, then, on Chalmers's view, assertions have both primary and secondary intensions, and these intensions cannot be read off the sentence asserted, but depend on the context including the a priori connections that the speaker associates with the words used.

Having given these two clarifications, I turn now to consider how Chalmers's two-dimensional approach might help users of the credence framework accommodate the fact that credence claims are opaque. We can begin with a central question for Chalmers's account: what are propositions? Are they primary intensions, secondary intensions, some combination of the two, or something else? Chalmers writes: 'It is natural for a two-dimensionalist to be a semantic pluralist, holding that there are many ways to associate expressions and utterances with quasi-semantic values, where different quasi-semantic values play different explanatory roles' (Chalmers, 2011a, p. 599). The thought is that we expect propositions to play many different roles (we expect them to be the objects of propositional attitudes, to be the content of declarative utterances, and to be truth-apt), and we cannot expect either primary intensions, secondary intensions, or some combination to fulfil all these roles—though perhaps they cover these roles between them. If we do insist that a single sort of thing—a proposition—should cover all of these roles (as is certainly traditional), then Chalmers's recommendation is that we focus on *enriched propositions*. I turn to Chalmers's account of enriched propositions later in this chapter. For now, I focus on just one of the roles that we expect propositions to play—and that is to be the objects of credence. What, in Chalmers's account, should play this role?

152 STATES AS METAPHYSICALLY POSSIBLE WORLDS

Chalmers writes that there are reasons why it is 'very tempting to suggest that primary intensions are the objects of credences' (Chalmers, 2011a, p. 618). This seems like a promising suggestion, and in the next section I explore this idea. Though the suggestion is promising, it raises many problems and questions. Chalmers raises some of these problems himself, pointing out that primary intensions cannot play all of the roles that we want the objects of credence to play (Chalmers, 2011a, p. 628), and suggesting some alternatives. We will come to these alternatives later in this chapter, but first I focus on exploring the initially promising suggestion that the objects of credence are primary intensions.

7.5 The objects of credence as primary intensions

On the view that we are exploring, the objects of credence are primary intensions, and a primary intension is a set of centred possible worlds. We can incorporate this view smoothly into the credence framework as follows:

- A state is a centred possible world: the set of all states (Ω) is simply the set of all centred possible worlds.
- We define an algebra over this set of states, and the resulting sets of states are the objects of credence.
- We can model the epistemic state of a rational agent with a credence function, obeying the probability axioms, which assigns some number between 0 and 1 to each element of the algebra.

So far the set-up seems fairly standard. Where is the innovation? The innovation is in how credence attribution statements work. The claim that the objects of credence are primary intensions rather than secondary intensions is (I argue) more or less empty unless taken as a claim about credence attribution statements. What would it mean to claim *simply* that the objects of credence are primary intensions, without stating or implying which assertions they are the primary intensions of? Primary intensions are (we are supposing—and admittedly there is some room for variation here) sets of centred metaphysically possible worlds, and secondary intensions are sets of metaphysically possible worlds. The claim that the objects of credence are primary intensions is not supposed to amount merely to the claim that they are sets of centred rather than uncentred metaphysically possible worlds. If the claim is to have significant content, then it must be taken to mean that in

THE OBJECTS OF CREDENCE AS PRIMARY INTENSIONS   153

each particular case the object of credence is the primary rather than secondary intension *of* some assertion—and the plausible candidate here is the assertion embedded in a credence attribution statement. The central claim, then, is that when we attribute a credence by giving an assertion, the object of credence is the primary intension of that assertion.

The mechanism by which we attribute credences might seem like something that should interest linguists rather than the wide range of people who use the credence framework, but in fact it is relevant for all users of the framework. This is not to claim that all users of the framework need to be experts in this field of course—but they do all need to make credence attribution statements (as they do whenever they give an agent's credence function or set out a decision table), and they need to understand these statements when made by others. We have seen in chapters 5 and 6 that there is room for error here, and we have seen some of the wide-reaching ramifications of such error, so understanding how credence attribution statements work is important for all users of the credence framework.

The key question is: when attributing credence to an individual, how do we designate the relevant object of credence—which in this chapter we are assuming is a set of (centred) metaphysically possible worlds? We cannot literally *point* at the set of worlds, because the only world that we can point at is the actual world.[7] So how can we go about designating the relevant set of possible worlds? If you think about how users of the credence framework usually do this, it is by giving an assertion. As an example, consider the top row of a decision table, where we designate each 'event' by putting an assertion in each cell (e.g. 'coin lands heads up', 'coin lands tails up'—or just 'H' and 'T' for short). How does such an assertion designate a set of possible worlds? The natural assumption is that the relevant set of possible worlds is the secondary intension of that assertion—that is, the set of worlds at which that assertion is true. But as we have seen, this leads to a problem: the secondary intension of 'George Orwell is a writer' is the same as the secondary intension of 'Eric Blair is a writer', but a rational agent can have a different credence in each. The view that we're currently considering seems to solve this problem, for we can instead take the relevant sets of possible

---

[7] Another way to designate a world is by describing it as the world closest to the actual world (if there is any one such world). We can also designate a set of worlds in this sort of way—e.g. the fifty closest worlds to the actual world (if there are fifty worlds that are closer than any others). But this is a very limited device for picking out worlds. Usually we designate possible worlds not just by their closeness to the actual world, but also (or instead) by what is true at that world. For example, we might designate a world as the closest world *where snow is not white*.

154 STATES AS METAPHYSICALLY POSSIBLE WORLDS

worlds to be the primary intensions of these assertions, and the primary intension of 'George Orwell is a writer' is different from the primary intension of 'Eric Blair is a writer'. We can give a similar account of a typical credence attribution statement that a user of the credence framework might make in the course of stating some agent's credence function:

(7e) S has a credence of v that P.

On the proposed view, (7e) states that S has a credence of v in the primary intension of P: that is, in the set of centred possible worlds at which P is verified (rather than true). This explains why we can truly say that Tom has a credence of 0.8 that George Orwell is a writer, and also truly say that Tom has a credence of 0.1 that Eric Blair is a writer, because though the secondary intension of 'George Orwell is a writer' is the same as the secondary intension of 'Eric Blair is a writer', the primary intensions of these assertions are different.

This looks like a promising way forward, then. But things are a bit more complicated than they appear. For recall that it is not sentence types but sentence tokens or assertions that have primary (and secondary) intensions. The primary intension of an assertion depends not just on the sentence type uttered, but on the context, including who is doing the uttering.[8] When we say that (7e) states that S has a credence of v in the primary intension of P— what primary intension is this? Is it the primary intension of 'P' were it to be uttered by the subject? Or the primary intension of 'P' as it has just been uttered by the attributer? Or what? To see a case where this might make a difference, recall Chalmers's case of Leverrier. For Leverrier, it is a priori that Neptune perturbs the orbit of Uranus, but this is not a priori for his wife. Suppose, then, that Leverrier says:

(7f) My wife's credence that Neptune perturbs the orbit of Uranus is 0.5.

---

[8] An interesting alternative (Bjerring, personal correspondence) is to claim—in opposition to Chalmers—that there is no variation in a priori connections amongst speakers. This would make primary intensions speaker-invariant, and so solve the problem discussed in the main text. It is not obvious, however, what follows a priori in this speaker-invariant sense: is it a priori that George Orwell wrote *1984*? Or that he took on the pen name of 'George Orwell'? There are rational agents who are uncertain whether George Orwell wrote *1984*, and who do not know whether he was ever called 'George Orwell' in his lifetime, and one worry is that in order to accommodate such agents we will be forced to hold that the name 'George Orwell' comes with so impoverished a set of a priori connections that the resulting primary intensions are not plausibly the objects of credence.

THE OBJECTS OF CREDENCE AS PRIMARY INTENSIONS   155

Here (on the view that we are considering) Leverrier is saying that his wife has a credence of 0.5 in some primary intension—but what primary intension? Presumably it is the primary intension of 'Neptune perturbs the orbit of Uranus'—but what the primary intension of this sentence is depends on who utters it. Is the relevant primary intension the primary intension it would have were his wife to utter it, or the primary intension it has as he (Leverrier) utters it?[9] If we take it to be the primary intension it has as he (Leverrier) utters it, then we have a problem. For the primary intension of 'Neptune perturbs the orbit of Uranus' as Leverrier utters it is the set of all centred possible worlds, for 'Neptune perturbs the orbit of Uranus' is a priori for Leverrier. So Leverrier would be stating that his wife has a credence of 0.5 in the set of all centred possible worlds, which entails that his wife is irrational. But this doesn't seem right: Leverrier's statement (7f) may be true, even though his wife is perfectly rational.

This is a symptom of a general problem. If we take the relevant primary intension in statements of the form (7e) to be the primary intension that P has in the mouth of the attributer, then we get a new version of the problem that drove us away from the standard account of propositions in the first place. The original problem was that when some assertion P is metaphysically necessary, then whenever a speaker attributes to someone a credence of less than 1 in P it follows from that attribution that the subject is irrational. Now we have a new version of that problem: whenever some assertion P is a priori for the speaker, then whenever the speaker attributes to someone a credence of less than 1 in P, it follows from the attribution that the subject is irrational. For example, if it is a priori for me that George Orwell is Eric Blair, and I say that you have a credence of less than 1 that George Orwell is Eric Blair, it will follow from my attribution that you are irrational. In general, then, for the decision theorist it will be difficult to express the credence state of any agent who does not share the decision maker's a priori knowledge without immediately classing that agent as irrational.

---

[9] A third possibility, suggested by a reviewer, is that in a sense statement (7f) might involve *both* primary intensions of the assertion—that is, both the primary intension that it would have if uttered by Leverrier, and the primary intension that it would have if uttered by his wife. On one version of this idea, statement (7f) should be understood as involving a quantifier: perhaps (7f) states that at least one of the two relevant primary intensions is such that Leverrier's wife has a credence of 0.5 in it. On another version of this idea, statement (7f) is ambiguous between the two possible primary intensions. These are intriguing suggestions, and (though I don't attempt that here) it would be interesting to trace the implications of each for the probability axioms and for the deference principles discussed below.

156 STATES AS METAPHYSICALLY POSSIBLE WORLDS

Let us, then, instead try saying that in (7f) the relevant primary intension is the primary intension of 'Neptune perturbs the orbit of Uranus' were Leverrier's wife to utter those words. I think that this is a more promising approach, although (as we shall see) it faces some serious problems. An initial problem is that it doesn't seem to fit well with our everyday use of credence attributions.[10] To see this, suppose that I hear someone give a great job talk on counterfactuals, and come to have a very high credence that the person will get the job—and say so to other people in the department. I arrived too late to hear the name of the person giving the job talk but let's say that in fact she is called 'Amanda Barnes'. It seems, then, that someone who knows her name can truly say:

(7g) Anna has a very high credence that Amanda Barnes will get the job.

(7g) states that I have a very high credence in some object: what object is that? The view that we are currently considering is that the relevant object is the primary intension of 'Amanda Barnes will get the job' as uttered by me—Anna (the subject). But what is the primary intension of this assertion as uttered by me? I would not utter this sentence, for I have never heard the name 'Amanda Barnes'. I associate no particular a priori connections with this name, and so it is quite unclear in which centred possible worlds the assertion is verified.

As a theory of how we currently use credence attribution statements, then, this is a poor theory. But perhaps as users of the credence framework this need not worry us too much, for it is not our job to chart actual linguistic practice. We need to be able to make and understand credence attribution statements, for this is exactly what we do when we put forward and assess descriptions of an agent's epistemic state, but we can be revisionary about the practice: instead of giving an account of how credence attributions currently work, we can instead stipulate a new convention for attributing credences. As we shall see, however, there are problems with this new convention.

---

[10] Another way to demonstrate the same point involves indexicals: for example, the sentence 'S has a high credence that I am hungry' is *not* typically taken to mean that S has a high credence in the primary intension of 'I am hungry' were S to utter those words. For a related discussion, see (Weber, 2013).

7.6 A new convention for credence attribution statements

On the view that we're considering, when attributing to someone (S) a credence of v in P, the relevant object of credence is the primary intension of P *as uttered by the subject (S)*. This is certainly not standard linguistic practice, but we might adopt it as a new convention on the grounds that users of the credence framework are not in the business of charting current linguistic practice, but just need a way of attributing credences. There are problems with this proposal, however, and the new convention cannot be adopted without repercussions for users of the credence framework. I explain these problems in this section.

The source of the problems is that our new convention disrupts a natural assumption made by users of the credence framework: that the objects of credence are common property, and that it is a straightforward matter to state that two different people have credences in the very same object of credence, or that one agent has credences in the same object of credence at two different times. For example, we might say that my credence that grass is green is lower than yours; or that my credence that grass is green has dropped over time. But the new convention that we are considering means that these sorts of statements are more complex than they first appear.

Consider these two statements:

(7a)   Tom's credence that George Orwell is a writer is 0.8.

(7h)   Tess's credence that George Orwell is a writer is 0.5.

From these two statements we might naturally infer that Tom has a higher credence than Tess in some proposition—namely that George Orwell is a writer. But given our new convention, that inference no longer follows. Statement (7a) attributes to Tom a credence of 0.8 in the primary intension of 'George Orwell is a writer' as uttered by Tom, and statement (7h) attributes to Tess a credence of 0.5 in the primary intension of 'George Orwell is a writer' as uttered by Tess. And the primary intension of 'George Orwell is a writer' might be different depending on whether it is uttered by Tom or Tess because they may associate different a priori connections with the name. For Tom perhaps 'George Orwell is a writer' is verified at all worlds where the person who created *Animal Farm* is a writer; while for Tess perhaps 'George Orwell is a writer' is verified at all worlds where the person whose portrait hangs in her university library

158 STATES AS METAPHYSICALLY POSSIBLE WORLDS

above the name 'George Orwell' is a writer. And the set of worlds where the person who created *Animal Farm* is a writer is different from the set of worlds where the person whose portrait hangs in Tess's university library above the name 'George Orwell' is a writer. Thus (7a) states that Tom has a credence of 0.8 in some primary intension, and (7h) states that Tess has a credence of 0.5 in some other primary intension, and so it does not follow that there is any proposition in which Tom's credence is higher than Tess's. More generally, on the proposed convention it is hard to tell whether two credence attribution statements relate to the same object of credence.

This causes problems.[11] One such problem concerns deference principles. To illustrate, suppose that Tom defers to Tess: that is, Tom regards Tess (so designated) as an expert. Then—given the standard definition of deference—for any proposition P and value v such that $Cr_{Tom}(Cr_{Tess}(P) = v) > 0$:

Deference: $Cr_{Tom}(P|Cr_{Tess}(P) = v) = v$

But $Cr_{Tom}(P) = v$ is a credence attribution statement, which (on the view that we're considering) attributes to Tom a credence of v in the primary intension of P as uttered by Tom. $Cr_{Tess}(P) = v$, in contrast, attributes to Tess a credence of v in the primary intension of P as uttered by Tess—which may be quite a different object. Why should Tom's credence in one object be constrained by Tess's credence in some other object? To illustrate the issue, let's suppose that Tom comes to have a credence of 1 that $Cr_{Tess}(P) = v$: that is, Tom has become certain that Tess's credence in some primary intension (the primary intension of P as uttered by Tess) is v. If Tom defers to Tess, then it follows given the standard definition of deference that $Cr_{Tom}(P) = v$: that is, Tom has a credence of v in the primary intension of P as uttered by Tom. But why should Tom match his credence in this object to Tess's credence in some other object? This seems like a strange thing for deference to require. Intuitively what deference ought to require is that were Tom to learn that Tess has a particular credence in some object, then Tom should assign the same credence *to the same object*. The problem is that the standard definition of deference given above no longer expresses this, given the new convention that we are considering for interpreting credence attribution statements. So either the new convention should be dropped, or the standard definition of deference must be adjusted.

---

[11] I focus on the problems that it causes to the users of the credence framework, but there are numerous linguistic anomalies too.

A NEW CONVENTION FOR CREDENCE ATTRIBUTION STATEMENTS   159

As a minimal adjustment, let us start by introducing the following notation. We can say that the primary intension of some assertion P, as uttered by S, is given by $PI_S(P)$. We are currently considering an approach on which the objects of credence are primary intensions, and on which standardly in a credence attribution statement the object of credence is the primary intension of the relevant assertion as uttered by the subject. Thus standardly the statement $Cr_S(P) = v$ means that S has a credence of v in the primary intension of assertion P as uttered by S: or—using our new notation—$Cr_S(PI_S(P)) = v$. Given this convention, above we interpreted the deference principle as though, where Tom defers to Tess, the following holds: $Cr_{Tom}(PI_{Tom}(P)|Cr_{Tess}(PI_{Tess}(P)) = v) = v$. This led to the strange result that Tom is required to align his credence in one object $(PI_{Tom}(P))$ with Tess's credence in a different object $(PI_{Tess}(P))$. To remedy this, we might maintain our new convention in general, but adjust the deference principle to force the objects of credence to be the same throughout. Thus we might say that where one person S_1 defers to another person S_2, then for any proposition P, value v, and possible person S_x, such that $Cr_{S1}(Cr_{S2}(PI_{Sx}(P)) = v) > 0$:

New Deference: $Cr_{S1}(PI_{Sx}(P)|Cr_{S2}(PI_{Sx}(P)) = v) = v$

This definition is in the spirit of the idea underlying the deference principle, and—given the view that we are considering—it seems like this principle (or something like it) is the best version of the deference principle that we can give. But there are repercussions for users of the credence framework: inferences that we might have expected to follow from the original deference principle do not go through on this new version, and furthermore it turns out to be very hard to conform to the new principle.

To see this, suppose (for the sake of argument) that Tom defers to Tess who is an expert on George Orwell. For Tess it is a priori that George Orwell was born in 1903 in India and was christened 'Eric Blair', though Tom has no idea about these biographical details. Tom learns that Tess has a credence of 0.8 that George Orwell died of tuberculosis. Given that Tom defers to Tess, we might, then, expect Tom to also have a credence of 0.8 that George Orwell died of tuberculosis—but the new principle of deference does not require this. What deference does require of Tom is that he should assign a credence of 0.8 to the primary intension *as expressed by Tess* of the assertion that George Orwell died of tuberculosis. Given the a priori connections that Tess associates with the name 'George Orwell', this is the set of centred

160   STATES AS METAPHYSICALLY POSSIBLE WORLDS

worlds where the person who was born in 1903 in India and christened 'Eric Blair' died of tuberculosis. But given that Tom does not know what a priori connections Tess associates with the name 'George Orwell', Tom does not know which is the relevant primary intension, and so it does not seem possible for him to assign it a credence of 0.8. Thus our new principle of deference is not only missing the implications that we might have expected, but it also seems to make it very hard for one agent to qualify as deferring to another.

The problems with the principle of deference also arise in cases of peer disagreement and conditionalization. In cases of peer disagreement, we have two individuals (Tom and Tim) who consider each other as peers, and yet disagree: that is, there is some P such that Tom's credence in P is v, Tim's credence in P is w, and $v \neq w$. The literature in this area debates how Tom and Tim should respond on learning of their disagreement. But the new convention that we are considering for credence attribution seems to undermine this debate. To say that Tom's credence in P is v is to say that Tom has a credence of v in the primary intension of P *as uttered by Tom*; and to say that Tim's credence in P is w is to say that Tim has a credence of w in the primary intension of P *as uttered by Tim*. Thus the objects of credence here may be different—in which case Tom and Tim are not disagreeing after all. Given the new convention that we are considering, then, the literature on disagreement will need to be rethought.

A further problem—with even deeper repercussions for users of the credence framework—concerns conditionalization. Conditionalization is a central principle of the framework: the idea is that your credence in P at t_1 after learning (just) some piece of evidence E should rationally match your credence at t_0 (before learning E) in P conditional on E. But given our new convention, this way of putting the principle is problematic. For your a priori connections can change over time, in which case the primary intension of P as uttered by you at one time can be different from the primary intension of P as uttered by you at a different time. Thus to say that you have a certain conditional credence at t_0 in P is to attribute to you a conditional credence in the primary intension of P as uttered by you at t_0, while to say that you have a certain credence at t_1 in P is to attribute to you a particular credence in the—possibly different—primary intension of P as uttered by you at t_1. And conditionalization should not require your credence at t_1 in some object to be constrained by your conditional credence at t_0 in some other object.

To illustrate, suppose that at time t_0 you associate the following a priori connection with the name 'George Orwell': George Orwell is the person whose entertaining photograph hangs in the department's library above the name 'George Orwell'. At this point, we can truthfully attribute to you a high credence that George Orwell is funny: that is, you have a high credence in the primary intension of 'George Orwell is funny' as uttered by you at t_0. You then come across several books by George Orwell and read them, gaining plenty of evidence in the process. We can now truthfully attribute to you a low credence that George Orwell is funny: that is, you have a low credence in the primary intension of 'George Orwell is funny' as uttered by you at t_1. On the view we are considering, however, it is not as though your credence in the same object has changed over time, for after reading George Orwell's books you presumably have new a priori connections (that George Orwell wrote *Animal Farm*, and so on), and perhaps it is no longer a priori for you that the portrait in your department's library is of George Orwell (perhaps having read his books you find it so hard to match up the author with the face that you come to doubt that the photograph has been correctly labelled). Thus your credence at t_1 that George Orwell is funny, and your credence at t_0 that George Orwell is funny are credences in two different primary intensions, and so it would be strange for the principle of conditionalization to require your credence at t_1 in one object to be constrained by your conditional credence at t_0 in some quite other object. In response, we would need to rethink the way that conditionalization is defined—perhaps along similar lines to the new definition of deference above.

In this section I've considered an account on which the objects of credence are primary intensions. At first this looked like a promising approach, for a primary intension is a set of centred possible worlds—just the sort of thing that we assumed an object of credence would be—and so it looked as though this account of the objects of credence could be smoothly incorporated into the credence framework. The innovation is in the way that we pick out an object of credence—as the primary (rather than secondary) intension of an assertion. But the primary intension of an assertion depends on who utters it and the utterer's a priori connections. Thus when we use an assertion to pick out an object of credence (which is what we do in credence attribution statements) we need to be clear on who is supposedly making the assertion. I found that if we supposed it was the attributer making the assertion, then we got a new version of the very problem that led us to explore the options that I am considering in this chapter. If we supposed instead that it was the subject making the assertion, then we faced numerous problems making sense of the definitions of deference principles and conditionalization.

162 STATES AS METAPHYSICALLY POSSIBLE WORLDS

I finish this section by describing one other way that we might interpret credence attribution statements, while maintaining that the objects of belief are primary intensions. Let us return to our general form of a credence attribution statement:

(7e) S has a credence of v that P.

We might maintain that statement (7e) attributes to S a credence of v in some primary intension, but deny that this is the primary intension of 'P' as asserted by either the subject or the attributer. On this view, the expression 'P' that appears in this statement is some sort of clue to what this primary intension is, but there is no straightforward way to decipher this clue to extract the relevant primary intension. I come back to this view later as it is related to Chalmers's own idea of 'appropriate' intensions (Chalmers, 2011a, p. 606). I just note here that—if this is the best view available—then it will have repercussions that users of the credence framework should take into account. It is not—as might be assumed—a straightforward matter to pick out some object of credence and talk about the credence assigned to it by some agent. The best that can be done is to mention an assertion, and thereby give some sort of indication of the object of credence picked out: that object is a set of centred possible worlds, and it is some sort of primary intension of the assertion made, but not necessarily the primary intension of the assertion as made by either the attributer or the subject.

Despite its initial promise then, the claim that the objects of credence are primary intensions faces problems however it is spelt out. I turn, then, in the next section to a different way we might use Chalmers's two-dimensional framework to understand credence claims.

7.7 The objects of credence as primary intensions and secondary intensions

The objects of credence cannot simply be secondary intensions: this is for the now familiar reason that an agent might have two different credences in the same secondary intension. We have also seen numerous difficulties for the view that the objects of credence are simply primary intensions. In this section I explore another possibility, which is that each object of credence is an ordered pair consisting of a primary intension *and* a secondary intension.

THE OBJECTS OF CREDENCE AS PRIMARY AND SECONDARY INTENSION 163

This view will involve a radical rethink of the credence framework, and here I just gesture towards one way that this account might work:

- There are two sets of states. A set of metaphysically possible worlds, and a set of centred metaphysically possible worlds. The set of metaphysically possible worlds are exclusive and exhaustive, and the set of centred metaphysically possible worlds are also exclusive and exhaustive.
- An object of credence is an ordered pair. The pair consists of a set of metaphysically possible worlds (a secondary intension), and a set of centred metaphysically possible worlds (a primary intension).
- We can no longer straightforwardly say that the domain of the credence function is an algebra over the set of states. For we have two sets of states, and the objects of credence are ordered pairs. How might we extend the idea of an algebra to cover this sort of case? What might it mean for the domain of the credence function to be closed under complement and union? Indeed, what is the complement of a given object of credence? To illustrate the issues here, suppose that our model has the following centred worlds $\{x, y, z\}$, and the following uncentred possible worlds $\{l, m, n\}$. And suppose that one element in the domain of the credence function is the following ordered pair: $<\{x\}, \{l\}>$. What is the complement of this object? Should we say that the complement is $<\{y, z\}, \{l\}>$? Should we instead (or perhaps, also?) say that the complement is $<\{y, z\}, \{m, n\}>$? There are no obvious answers to these questions.

The implications of all of this would require a great deal of thinking through, and I don't attempt to do that here.[12] Part of my reason for not making the attempt is that this account seems to be vulnerable to all the problems that we encountered in the previous section. Consider the following claim:

(7i)   Tom has a credence of 0.05 that George Orwell is Eric Blair.

This statement attributes to Tom a credence of 0.05 in some object of credence, and on the account that we're currently considering this object

---

[12] For those interested in exploring this idea further, Ed Elliot (personal correspondence) has suggested one way that this model might be simplified. The key idea is that the secondary intension of a proposition is determined by its primary intension together with an assignment of actuality, and this may allow us to simplify the algebra.

164 STATES AS METAPHYSICALLY POSSIBLE WORLDS

of credence is an ordered pair consisting of a primary and secondary intension. The secondary intension is the set of all metaphysically possible worlds. What is the primary intension? If it is the primary intension of 'George Orwell is Eric Blair' as uttered by the attributer, then it may be that the attributer's a priori connections with George Orwell are such that the primary intension of 'George Orwell is Eric Blair' is the set of all centred possible worlds. In this case (7i) states that Tom has a credence of 0.05 in the ordered pair consisting of the set of all metaphysically possible worlds, and the set of all centred possible worlds. Presumably (depending how the details of the account are spelt out) Tom will thus be classed as irrational. We might instead try stating that the relevant primary intension is the primary intension of 'George Orwell is Eric Blair' as uttered by the subject, but then the problems with deference and conditionalization principles discussed in the last section resurface.

We might try saying that an object of credence is an ordered pair consisting of the secondary intension, plus a primary intension that may be neither the primary intension of the relevant sentence as uttered by the subject nor as uttered by the attributer. This is related to an account of belief attribution that is discussed (though not ultimately endorsed) by Chalmers. He writes:

'S believes that P' is true of a given person iff that person has a belief with the secondary intension of P (in the mouth of the ascriber) and with a P-appropriate primary intension. (Chalmers, 2011a, p. 606)[13]

What exactly is it for a primary intension to be 'P-appropriate'? Chalmers states that 'no such complete account [of what it is for a primary intension to count as P-appropriate in a context] is yet close at hand, and because our judgments about attitude ascriptions are so unruly, comprehensive principles may be hard to find' (Chalmers, 2011a, p. 609). Chalmers goes on to examine our intuitions in a range of cases to build up some evidence that such an account would need to accommodate, but no complete account emerges. Dialectically, this is not an objection to Chalmers's theory in itself, for as far as I know no theorist has put forward a complete account of attitude ascriptions that tracks actual linguistic use. What interests us here, though, is whether an analogue of Chalmers's view might be adopted by

---

[13] I have changed the labels ('S', 'P') for consistency with the rest of this chapter.

THE OBJECTS OF CREDENCE AS PRIMARY AND SECONDARY INTENSION   165

users of the credence framework, who may have naturally expected that when they attempt to pick out an object of credence it should quite clear which object has been picked out.

Perhaps users of the credence framework will just have to be left dissatisfied, at least for the present. To state an agent's credence function involves giving the relevant objects of credence, which can only really be done using assertions. How does an assertion pick out the object of credence? If the object of credence is the *secondary* intension of the assertion, then we face the now familiar problem with opacity. If the object of credence is the *primary* intension of the assertion, then we face the challenges described in this chapter: if it is the primary intension of the assertion as uttered by the attributer (e.g. the user of the framework—the person writing the credence function up on the whiteboard), then we face a version of the opacity problem afresh; if it is the primary intension of the assertion as uttered by the subject, then we face problems with the idea of shared content which infect deference principles and conditionalization. The remaining alternative is to say that the object picked out is some 'appropriate' primary intension, and then we are left somewhat in the dark as to which set of worlds is intended. Not entirely in the dark: this is a research project in progress, and some rules to constrain the choices are available. But there is as yet no clear and universally agreed answer—and users of the credence framework would be wise to bear this in mind and set out and interpret credence claims and decision tables with caution. These points apply with just as much force should we claim that an object of credence consists of both a primary intension *and* a secondary intension: this raises numerous technical challenges for the credence framework, and the difficulty in identifying the relevant primary intension remains.

On a practical note, users of the credence framework might respond by taking great care when choosing an assertion to pick out an object of credence. If an assertion is made using entirely neutral vocabulary—no indexicals (including the hidden indexicals to be found in tensed assertions), no demonstratives, no proper names, and no natural kind terms—then the primary intension of the assertion is simply identical to its secondary intension. If we were to limit ourselves, then, to assertions made purely in neutral vocabulary, then the issues raised in this chapter would not arise, and indeed the problem of opacity in general would vanish. But there are many problems with carrying this out in practice. Our ordinary language terms are suffused with non-neutral vocabulary, and it is quite impractical to expect users of the credence framework to avoid it. Furthermore, users of the

166 STATES AS METAPHYSICALLY POSSIBLE WORLDS

framework may wish to talk about issues that can arguably only be expressed when non-neutral vocabulary is allowed. For example, a welfare economist may question whether it is permissible for me to save *myself* rather than another—and here the indexical seems essential to the question. Rather than attempting to eliminate non-neutral vocabulary, I would recommend that users of the credence framework should be aware of the issues and the possible implications—and this book is largely about motivating this alert stance.

In the final section of this chapter, I turn to consider one more way in which we might use the two-dimensional framework to give an account of the objects of credence.

7.8 The objects of credence as enriched propositions

In the previous section I tried taking an object of credence to be an ordered pair, consisting of a primary and secondary intension. This idea relates to a view of belief attribution discussed but not ultimately endorsed by Chalmers. In this section I briefly consider whether the objects of credence might instead be taken to be the structured entities that Chalmers calls 'enriched propositions'. Chalmers suggests that these are the best candidates for the objects of belief (Chalmers, 2011a, p. 633), so the hope is that they will also prove good candidates for the objects of credence.

I briefly introduced structured propositions in chapter 2. The general idea is that a proposition consists of various parts (typically these parts are taken to be objects and properties) structured in a particular way. In that chapter I also introduced possible world semantics. According to possible world semantics, an assertion has both an extension (a truth value at the actual world) and an intension (a function from possible worlds to truth values). In addition, the parts of the sentence asserted—the names and predicates—also have both an extension and an intension: the extension of a name is the object that it designates, and the intension of a name is a function from possible worlds to objects; the extension of a predicate is the set of objects in the actual world that fall under it, and the intension of a predicate is a function from possible worlds to sets of objects. Chalmers's view of enriched propositions incorporates elements of both of these views.

On Chalmers's view, an assertion has both a primary and a secondary intension. And the parts of the sentence asserted—the names and predicates—also have both primary and secondary intensions. The secondary intension of 'George Orwell', for example, is a function from possible worlds to

THE OBJECTS OF CREDENCE AS ENRICHED PROPOSITIONS 167

objects: this function will map every world (at which George Orwell exists) onto the same person. The primary intension of 'George Orwell' is a function from centred possible worlds to objects: this function will map the various centred possible worlds onto a range of different people. To see which object a given centred possible world will be mapped onto, we consider the hypothesis that the centred possible world (described using its canonical specification) is actual, and then find an object such that it is a priori that that object is George Orwell given our hypothesis. For example, take our centred possible world w* in which the person who was christened 'Eric Blair' never wrote any books, and someone else took the pen name of 'George Orwell' and wrote *1984, Animal Farm*, and so on. Under the hypothesis that this is how things actually are, it is a priori that George Orwell is the person who wrote *1984* and so on. Thus the primary intension of 'George Orwell' maps w* to this person.

Thus each expression can be associated with both a primary and a secondary intension. Chalmers defines an expression's 'enriched intension' as the ordered pair consisting of that expression's primary and secondary intensions. From here Chalmers gives his definition of an 'enriched proposition': 'The enriched intension of a complex expression is a structure consisting of the enriched intension of its simple parts (including any unpronounced constituents), structured according to the expression's logical form. The enriched intension of a sentence [Chalmers clarifies elsewhere that by this he means a sentence in context, i.e. an assertion] is its associated enriched proposition' (Chalmers, 2011a, p. 600). Chalmers comes close to claiming that this is his account of what propositions are (Chalmers, 2011a, p. 603). And it is natural to assume that the objects of credence are propositions—and so we arrive at the view that the objects of credence are enriched propositions.

Suppose, then, that we were to accept that the objects of credence are enriched propositions—the structured entities as described. On this view, if objects of credence could be described as sets at all, then the elements of these sets would clearly not be simple metaphysically possible worlds. If we were to incorporate this view into the credence framework, on which the objects of credence are sets of states, then we would need to drop the claim that states are metaphysically possible worlds—or entirely rethink the credence framework. Thus whatever the merits of this proposal, it is clearly a strategy that goes beyond the remit of this chapter.[14]

---

[14] A theorist may maintain that (enriched) propositions are structured entities, but that the objects of credence are not themselves structured entities: rather, the objects of credence are sets of metaphysically possible worlds that are determined by the relevant enriched propositions

168 STATES AS METAPHYSICALLY POSSIBLE WORLDS

7.9 Chapter summary

On a natural interpretation of the credence framework, states are metaphysically possible worlds. Given some apparently obvious assumptions, this natural interpretation of the credence framework faces a problem given that credence claims are opaque. There are broadly two ways that users of the credence framework might respond: they might drop the claim that states are metaphysically possible worlds, or they might question one of the apparently obvious assumptions. In this chapter I have focused on this second strategy. I drew on Russell's descriptivism (7.2), Stalnaker's account of belief attribution (7.3), and Chalmers's two-dimensionalism (7.4—7.8), but the strategy led to numerous problems. I turn, then, in the next chapter to try the alternative strategy, which is to drop the claim that states are metaphysically possible worlds.

(Chalmers, personal correspondence). This may be a plausible position—my aim in this section was only to explore the rather different view that the objects of credence themselves might be enriched propositions.

8

States as Something Else

8.1 Introduction

In this part of the book, I'm focusing on foundations: how can the credence framework accommodate our tenet that credence claims are opaque? We saw that if we spell out the details of the credence framework in a natural way—with states interpreted as metaphysically possible worlds, and various other natural assumptions in place—the credence framework is unable to accommodate our tenet. In the last chapter (7) I tried rejecting some of the 'other natural assumptions', but that proved unsuccessful. In this chapter I try instead dropping the idea that states are metaphysically possible worlds.

We can see how this strategy might solve our problem. Each object of credence—that is, each proposition—corresponds to a set of states, and these states need not be metaphysically possible worlds. Consider, then, the two credence attribution statements (8a) and (8b) below:

(8a) Tom has a credence of 0.8 that George Orwell is a writer.

(8b) Tom has a credence of 0.1 that Eric Blair is a writer.

These two sentences attribute certain credences to Tom. Sentence (8a) attributes to Tom a credence of 0.8 in the proposition expressed by 'George Orwell is a writer', and this proposition corresponds to the set of states at which George Orwell is a writer. Sentence (8b) attributes to Tom a credence of 0.1 in the proposition expressed by 'Eric Blair is a writer', and this proposition corresponds to the set of states at which Eric Blair is a writer. If there are states where George Orwell is a writer and Eric Blair is not, or vice versa, then the proposition expressed by 'George Orwell is a writer' is different from the proposition expressed by 'Eric Blair is a writer', and so it is no longer a problem for Tom to have different credences in each, and (8a) and (8b) can both be true, as desired.

This, then, looks like a promising strategy. Indeed, when I first recognized that credence claims are opaque, and realized that this creates a problem

The Objects of Credence. Anna Mahtani, Oxford University Press. © Anna Mahtani 2024.
DOI: 10.1093/oso/9780198847892.003.0008

170 STATES AS SOMETHING ELSE

given a natural way of interpreting the credence framework, the idea of taking states to be something other than metaphysically possible worlds seemed like the obvious solution. Surely all we have to do (I thought) is interpret states as more fine-grained than metaphysically possible worlds! I still think that this strategy has promise, but it comes with various complications and challenges that I lay out below. A first challenge is to figure out what these more fine-grained states could be, and here I draw on the literature on impossible worlds (Priest, 2005; Nolan, 2013; Jago, 2014; Berto and Jago, 2019) and show how we naturally reach the conclusion that states correspond to sets of sentences. But sentences or sentence-context pairs? In what language? And should each set of sentences be complete and coherent? In what sense? All these details need to be spelt out and their implications for the credence framework thought through if the strategy explored in this chapter is to be successful.

8.2 Fine-grained worlds

Let us begin with this question: if states are something other than meta-physically possible worlds, then what are they?[1] A natural thought is that these other states are just like metaphysically possible worlds, only more fine-grained. But when we try to describe these more fine-grained worlds, we seem forced to claim that they must be—at least partly—linguistic representations.[2]

To see why, consider that to accommodate the fact that credence claims are opaque, we need there to be a fine-grained world where George Orwell and Eric Blair are two different people, and where George Orwell is a writer but Eric Blair is not. What will this world be like? Perhaps we could describe

---

[1] Kit Fine states: 'Philosophers have been intrigued by the ontological status of impossible worlds. Do they exist and, if they do exist, then do they have the same status as possible worlds? To my own mind, these questions are of peripheral interest. The central question is whether impossible worlds or the like are of any use, especially for the purposes of semantic enquiry' (Fine, 2021, p. 144). Following Fine, I do not delve deeply into metaphysical questions but focus on those questions about the nature of states that are important to users of the credence framework.

[2] Or perhaps rather than a linguistic representation, we might take a possible world to be itself a proposition. If we combine this with the view on which a proposition is a set of possible worlds, then it seems that no non-circular definition of 'proposition' is available. This is not necessarily a problem, for we might take the view that propositions are basic and need not be explained in any other terms. A view rather like this can be found in (Prior and Fine, 1977), though I do not explore it further here.

FINE-GRAINED WORLDS 171

the world as follows: two different children are born; one is christened 'Eric Blair', and never writes a single book; the other eventually takes the pen name 'George Orwell' and writes *Animal Farm*, *1984*, and so on. But this could be the description of an ordinary metaphysically possible world. And at all metaphysically possible worlds, at least on the standard view of truth across such worlds, George Orwell *is* Eric Blair: the world as described is just a world where George Orwell/Eric Blair never did become a writer, and someone else took up the pen name 'George Orwell', and wrote *Animal Farm* and *1984* and so on.[3] We can, of course, take a non-standard view, and say that at this metaphysically possible world, George Orwell is not Eric Blair, but this is the sort of approach that we considered in the last chapter where we tried maintaining that states are metaphysically possible worlds and questioning some other natural assumptions—including the standard view of truth across metaphysically possible worlds. In this chapter we're supposed to be leaving that approach behind, and bringing more fine-grained worlds into the picture to resolve our problem, and these more fine-grained worlds are supposed to be different from metaphysically possible worlds. How are they going to be different?

The problem is that, in a sense, for every way that things could be, there is already a metaphysically possible world where things are that way. For example, consider all the ways that matter could be arranged: we might specify exactly which sorts of atoms of matter are to be placed in each location at every moment throughout all time. For any such arrangement of matter, there is a metaphysically possible world where matter is so arranged. There are limits to what is possible, of course—we cannot have two atoms occupying exactly the same location at the same time—but within such limits every combination is possible.[4] Call the complete arrangement of matter across all time at a world the microphysical facts at that world. Could there be two worlds that differ in some way, but share all their microphysical facts?[5] You might think that two worlds could differ with respect to qualia (that is, subjective experience) even though they share all

---

[3] Effectively this is world w^* as described in chapter 7, and we are 'considering it as counterfactual'.

[4] We might think that if we simply shift, rotate, or reflect the matter, the resulting world is not numerically distinct. A brief discussion of this sort of view can be found in (Lewis, 1986b), but I do not explore this question here.

[5] To put this question in terms of supervenience, it is helpful to define two senses of 'supervene': supervene$_M$ and supervene$_W$. Let us say that A facts supervene$_M$ on B facts iff there are no two metaphysically possible worlds differing over A facts but agreeing over B facts. And let us say that A facts supervene$_W$ on B facts iff there are no two worlds (which might be

172 STATES AS SOMETHING ELSE

their microphysical facts.[6] To accommodate this thought, we can consider all the ways that both matter and qualia can be arranged: we might specify exactly which sorts of atoms of matter are to be placed in each location at every moment throughout all time, and precisely what qualia (if any) is to be experienced by each object at each point. Within certain limits (e.g. no two atoms in the same location at the same time; arguably no object subject to two different qualia at the same time) every combination is possible: for any combination, there is a metaphysically possible world at which matter and qualia are so arranged. If we want worlds that are more fine-grained than metaphysically possible worlds, then any extra detail will have to be *in addition* to all microphysical and qualia facts. The extra detail will not arise from the microphysical or qualia facts: rather, once the microphysical and qualia facts about a world are fixed, the extra detail will need to be grafted on. How can we make sense of this idea?

We can find a model for this sort of idea by considering centred possible worlds. A centred possible world is an ordinary metaphysically possible world plus a centre, and a centre consists of a person and a time. Centred possible worlds are more fine-grained than metaphysically possible worlds: a single metaphysically possible world can be the base of many different centred possible worlds. Thus we can have two different centred possible worlds: one consisting of a particular metaphysically possible world, detailed down to the last atom and qualia, plus a centre consisting of a person and a time; the other consisting of the same particular metaphysically possible world, plus a centre consisting of a different person and time. Different facts hold at these two centred possible worlds. For example, it might be that at one it is true that I am hungry, but at the other it is false that I am hungry: the atoms and qualia are arranged identically at both worlds, but the worlds have different centres, and the centre of the first world is hungry whereas the centre of the second world is not. What is true at a centred possible world, then, does not depend just on the microphysical facts at that world. Nor is it fixed even by the microphysical facts together with the supplementary facts about qualia.

We can use an analogous idea to construct the sort of fine-grained worlds that we need to handle the fact that credence claims are opaque.

---

more fine-grained than metaphysically possible worlds) differing over A facts but agreeing over B facts. The question in the main text is effectively whether there are any facts that do not supervene$_W$ on microphysical facts.

[6] See, for example, (Jackson, 1982) for such a view.

FINE-GRAINED WORLDS   173

For fine-grained worlds to play the role that we need them to, we need a fine-grained world where George Orwell and Eric Blair are two different people. Perhaps we can just take an ordinary metaphysically possible world and graft this extra non-identity claim onto it. The non-identity claim will not be fixed by the microphysical or qualia facts at that world, but will rather be a 'further fact' that holds in that world.[7] To make this solution work quite generally, we can graft a set of identity and non-identity claims onto each metaphysically possible world: if George Orwell is not Eric Blair, then this claim will be included; if Mount Everest is Snowdon, then this will be included, and so on.[8]

As well as grafting on identity claims involving names, we would also need to graft on claims involving predicates. In this book when discussing opacity I have generally focused on proper names: the usual example I use to demonstrate that credence claims are opaque involves Tom's credences about George Orwell and Eric Blair. But there are also examples involving predicates. Take, for example, the two claims below:

(8c)   Tom's credence that there is water in his cup is 0.8.

(8d)   Tom's credence that there is H_2O in his cup is 0.1.

It seems that both (8c) and (8d) can be true, without Tom being irrational: Tom might not know that water is H_2O, in which case he might have a low credence that there is H_2O in his cup even while he has a high credence that there is water is his cup. If we took the objects of credence to be sets of metaphysically possible worlds, then we would have a problem. This is because 'water' is standardly taken to be a natural kind term, and natural kind terms are standardly taken to be rigid designators. On this view, water couldn't have been composed out of different chemicals, or it wouldn't have been water. Of course, it could have been that a liquid with a different chemical formula flows in our rivers and so on: there is a metaphysically possible world where that happens. Indeed, there is a metaphysically possible world which is exactly like ours, except that at this non-actual world there is a substance composed of something other than

---

[7] This is to embrace haecceitism, according to which there are primitive non-qualitative facts about identity. See (Forbes, 1985; Mackie, 1987) for an interesting debate over the possibility of 'bare identities'.

[8] See Kripke here for the claim that the identities of objects across possible worlds can be secured simply by stipulation (Kripke, 1980, p. 44).

174 STATES AS SOMETHING ELSE

H_2O which behaves exactly like water does in the actual world, and at this non-actual world people treat this substance just how we actually treat water—even calling it 'water'. But (on the standard view), the world described is not a metaphysically possible world where water is not H_2O. Rather, it is a metaphysically possible world where the stuff that is in the rivers and so on is not H_2O and so is not water. On the standard view, then, the set of metaphysically possible worlds where there is water in Tom's cup is exactly the same as the set of metaphysically possible worlds where there is H_2O in Tom's cup. The problem, then, is that (8c) and (8d) seem to be attributing different credences to Tom in the very same proposition.

If we want to resolve this problem by switching from metaphysically possible worlds to more fine-grained worlds, then we'll need a fine-grained world where Tom's cup contains water but doesn't contain H_2O. What will this world be like? We might think of a world where Tom's cup contains a liquid other than H_2O—a liquid that is odourless and colourless, essential to life, and the substance that fills all the rivers and oceans. But there is already an ordinary metaphysically possible world like that—and at that metaphysically possible world Tom's cup does not contain water. Once again, then, we'll just have to graft on a further fact: at this world, the liquid in Tom's cup is water, but not H_2O. More generally, then, in addition to our list of identity claims, we must also add a list of claims assigning (perhaps partial) extensions to various predicates. One such list will assign to 'water' an extension that includes the contents of Tom's cup, and assign to 'H_2O' an extension that does not include the contents of Tom's cup, and at this world Tom's cup will contain water but not H_2O.

On this view, then, we make worlds more fine-grained by taking ordinary metaphysically possible worlds, and grafting onto them a set of additional claims about identities and the extensions of predicates. So a fine-grained world will consist of an ordinary metaphysically possible world together with a set of additional claims. There are several reasons to be dissatisfied with this account. Firstly, it seems that in some cases a fine-grained world will consist of a metaphysically possible world at which certain facts obtain, together with additional claims that contradict some of those facts. To see this, consider that at every metaphysically possible world George Orwell is Eric Blair, and yet there will be a fine-grained world that consists of a metaphysically possible world, with the following claim grafted onto it: 'George Orwell is not Eric Blair'. Such a world is rather like René Magritte's painting *The Treachery of Images*, consisting of a painting of a pipe together with the words 'Ceci n'est pas une pipe' ('This is not a pipe').[9] Here we have a disanalogy with centred possible worlds. A centred possible world consists of an ordinary metaphysically possible world together with a centre, where that centre is a particular person and time. The claims that hold at that ordinary metaphysically possible world are not in conflict with the claim that the centre is a particular person and time: no particular facts about the centre hold at an ordinary metaphysically possible world, and so adding on a claim about the centre does not introduce conflict. Instead, adding on a claim about the centre can be seen as a way of fine-graining metaphysically possible worlds. Adding on claims involving identity and predicates, on the other hand, are not naturally seen as a way of fine-graining metaphysically possible worlds. For claims about identity and the extension of predicates already hold at ordinary metaphysically possible worlds, and so if the claims being grafted on are to make any difference at all, they will need to at least sometimes introduce conflict. This, then, is one reason to be dissatisfied with the idea of a fine-grained possible world as a metaphysically possible world with some additional facts grafted onto it.

Another reason to be dissatisfied with this idea is that it is unnecessarily complicated. Why have a fine-grained possible world be a metaphysically possible world with a collection of claims attached, when we might instead have it simply be a collection of claims? This idea—that a fine-grained possible world might be just a collection of claims—seems like a promising approach. In the next section I put this idea into context by describing a broad range of different ways that philosophers have thought about possible worlds.

8.3 Linguistic representations

Lewis categorizes views about possible worlds into two broad camps: views on which there are 'genuine' non-actual possible worlds; and views on which there are only 'ersatz' non-actual possible worlds. Those who hold that there are genuine non-actual possible worlds think that these worlds exist in much the same way that the actual world exists. Lewis held such a view about metaphysically possible worlds (Lewis, 1986b), but most theorists hold the

---

[9] Many thanks to Nicholas Makins for pointing out this connection.

176 STATES AS SOMETHING ELSE

less radical view that non-actual possible worlds are ersatz possible worlds, which are merely representations or models of how things could be, and for our purposes here I will also hold that non-actual possible worlds are ersatz possible worlds.[10] Within the category of ersatz possible worlds, Lewis further distinguishes two different sorts of representations or models: pictorial representations and linguistic representations (Lewis, 1986b).[11]

You might be used to thinking of possible worlds as pictorial representations. For example, suppose that someone asks you: could kangaroos be blue? Perhaps your immediate reaction is to imagine—picture 'in your mind's eye'—a world where kangaroos are blue, and then you will answer their question affirmatively. Of course, sometimes senses other than sight will be involved. For example, you might instead be called upon to imagine a world where dogs sound like cars when they move around, and then your imaginative model of the relevant world will come with sound effects too. The general idea is that often when we are thinking about a possible world, we summon up some sort of perceptual model. But if we want to think about worlds that are more fine-grained than metaphysically possible worlds, then can we think of them as pictorial representations? I don't think we can. For every possible combination of microphysical and qualia facts, there is already a metaphysically possible world that represents that combination. How can we summon up, in imagination, a world that is more fine-grained than any metaphysically possible world? By exploring such a world in your imagination, what could you hope to perceive that would not arise from the microphysical and qualia facts that hold at that world?[12] Thus worlds that are more fine-grained than metaphysically possible worlds cannot be—or at least cannot *just* be—pictorial representations.

This leaves us with the view on which worlds are linguistic rather than pictorial representations. Several such accounts are discussed in the literature. Richard Jeffrey, for example, takes worlds to be 'complete, consistent novels' (Jeffrey, 1965), and related views are discussed by (Carnap, 1947; Hintikka, 1962; Adams, 1974; Melia, 2001; Sider, 2002; Jago, 2014). Some theorists have produced accounts of worlds as linguistic representations as a

---

[10] Takashi Yagisawa (Yagisawa, 1988) argues for a very radical conclusion—that *impossible* worlds also exist, in the same sort of way that Lewis thinks that metaphysically possible worlds exist. I don't attempt to explore this interesting approach here.

[11] Lewis defines and dismisses a further category—'magical ersatzism'—recently defended by (Nolan, 2015). I don't include a discussion of this further category here.

[12] One answer is that you might perceive your own place in that world. To address this point, read 'metaphysically possible world' throughout this paragraph as 'centred metaphysically possible world' and add 'centre facts' to the list consisting of microphysical and qualia facts.

way of trying to solve the problem of logical omniscience—a problem that is rather different from the problem of opacity that we are grappling with.[13] In places, as we will see, this means that some of the accounts in the literature are not able to solve our problem. On the other hand, because we are not trying to solve the problem of logical omniscience, we can construct a less radical account of worlds than many of these theorists, and so avoid some of the objections raised against these accounts. I begin in the next section with an exploration of the basic idea of taking a world to be a linguistic representation.

8.4 Sentence-worlds

What does it mean to say that worlds are linguistic representations? We can start with the simplest idea—that a world is just any set of sentences. For example, we might say that the following is a world:

{'George Orwell is not Eric Blair'; 'grass is green'}

An immediate objection is that we normally think of worlds as being 'complete'—that is, as specifying every detail—while this world leaves a lot of details unspecified. Should we require that our sentence-worlds are complete? And what would this mean? Would it mean, for example, that each world ought to include every grammatical sentence in every language, unless it contains its negation? And should we also require that our sentence-worlds are coherent? What would that mean? Would it mean just that no world should contain both a sentence and its negation? Or should the coherence requirement be stronger? I return to these important questions in section 8.7. But first—as this book is primarily focused on the credence framework—I consider how credence claims are supposed to work given this idea of worlds as linguistic representations.

Take the following credence attribution statement:

(8e) Tom has a high credence that Paris is in France.

---

[13] The two problems are of course related. I say that they are different because the problem of opacity threatens to require omniscience quite generally, rather than mere logical omniscience. See section 4.4 for discussion on this point.

178 STATES AS SOMETHING ELSE

Here the expression 'Paris is in France' is directing us to the relevant object of credence, which is a set of states. And on the view that we're exploring in this chapter, a state (or world) is a linguistic representation: it is a set of sentences, which we can call a 'sentence-world'. Thus the object of credence is a set of such sentence-worlds—where each sentence-world is a set of sentences. How does the expression 'Paris is in France' pick out the relevant set of sentence-worlds? The natural thought is that the sentence-worlds are those that contain the sentence 'Paris is in France'.[14] The object of credence is the set of all and only those sentence-worlds that contain the sentence 'Paris is in France'.

But this presents us with a problem. For the word 'Paris'—like many words—is ambiguous. There is a Paris in France, and a Paris in Texas (and many other places are also called 'Paris'). Context is needed to fix which Paris is being referred to in claim (8e). Perhaps Tom is on a train trip around Europe. He sees a train headed for 'Paris', and decides to board it. He's almost sure that Paris is in France, but has a niggling doubt and thinks it might be in Belgium. A friend travelling with him on the train who knows his state of mind could truly utter (8e) under these circumstances. But suppose now that whilst on the train Tom receives an email inviting him to a conference in Paris, Texas. Tom is quite sure that this Paris—the one where the conference is—is not in France. He sends back an email asking whether his airfare will be covered, and saying that he has always wanted to visit the US. At the same time as Tom's companion utters (8e), and with the same 'Tom' in mind, the conference organizer can truly say:

(8f)   Tom has a low credence that Paris is in France.

In this way, (8e) and (8f) can both be true together, because they are uttered in different contexts, with 'Paris' referring to a different city in each. This is just a symptom of the familiar fact that the content (and truth-value) of an assertion depends not just on the sentence type uttered, but also on the context. But it creates a problem for the view on which the objects of credence are sentence-worlds. In both claims (8e) and (8f), the expression 'Paris is in France' is directing us to the object of credence, which is a set of

---

[14] If sentence-worlds are not complete—in that for some sentence neither it nor its negation is included in a given sentence-world—then we might want to say that the relevant sentence-worlds are not just those that contain the sentence 'Paris is in France', but also those whose set of sentences *imply* the sentence 'Paris is in France'. I return to this issue in section 8.7, but just note here that the issue discussed in the text arises either way.

sentence-worlds. If we assume that the relevant set of sentence-worlds include all and only those sentence-worlds that contain the sentence 'Paris is in France', then both (8e) and (8f) are directing us to the very same object of credence. Thus (8e) and (8f) are inconsistent: Tom cannot have both a high and low credence in the very same object of credence—at least not if his epistemic state can be represented by a credence *function*. We switched from metaphysically possible worlds to sentence-worlds because metaphysically possible worlds were too coarse-grained: it seemed that an agent could have different credences in the same set of metaphysically possible worlds. But we seem to have ended up with a similar problem to the one that we started with: we have two credence attribution statements, assigning different values to what—on our account—is classed as the very same proposition. It seems that sentence-worlds are not sufficiently fine-grained. How can we fine-grain them more? There are broadly two approaches that we might take, and I describe them below in sections 8.5 and 8.6.

8.5 Sentence-context-worlds

The reason why both (8e) and (8f) can be true is that they are uttered in different contexts. As discussed in chapter 2, the truth-value of an utterance does not just depend on the sentence type that is uttered, but also on the context in which it is uttered. This is particularly obvious when indexicals are involved (obviously 'I am hungry' might be true as uttered by me at a certain time, but false as uttered by someone else, or by me at a different time). But the same phenomenon holds for other terms too. Often the same proper name has multiple possible referents (as does 'Paris'), and context is needed to single out which particular object or individual is referred to.[15] Similarly many predicates are ambiguous: for example, the predicate ' . . . is a bank' can relate to river banks or money banks. Thus it is hardly surprising that (8e) and (8f) can both be true if they are uttered in different contexts.

This suggests that the objects of credence are not sets of sentence-worlds— where sentence-worlds are sets of sentences—but rather sets of what we might call 'sentence-context-worlds', where a sentence-context-world is a set of sentence-context pairs. One sentence-context pair consists of the sentence 'Paris is in France' together with the context in which Tom's

---

[15] In fact context has more to do here than just single out the object or individual referred to, as Kripke's Paderewski problem demonstrates (Kripke, 1979).

180 STATES AS SOMETHING ELSE

train companion makes his utterance; and another sentence-context pair consists of the sentence 'Paris is in France' together with the context in which the conference organizer makes his utterance. A sentence-context-world might contain the first of these but not the second, or vice versa. We might, then, say that (8e) as uttered by Tom's train companion is attributing a high credence to Tom in one object of credence, namely the set of all sentence-context-worlds containing the first sentence-context pair described above; whereas (8f) as uttered by the conference organizer is attributing a low credence to Tom in a different object of credence—namely the set of all sentence-context-worlds containing the second sentence-context pair described above. This would explain how both (8e) and (8f) can be true together.

But there is a problem with this suggestion. In the last section, when I was discussing sentence-worlds, I said that we might require a sentence-world to be complete, and it was easy enough to see at least one way that we might satisfy this requirement: we might require that a sentence-world contains every grammatical sentence in every language, unless it contains its negation. But how can we pull off the same trick with contexts? Various sentences have been uttered in various contexts, but obviously not all sentences have been uttered in all contexts. Some sentences have been uttered in some contexts but not in others, and some sentences have not been uttered at all. We might aim to stock our sentence-context-worlds with every possible combination. To visualize this, we can picture a matrix with a left-most column specifying every grammatical sentence, and a top-most row detailing every possible context: then each cell of this matrix will give us a sentence-context pair.[16] We can then rule that every such sentence-context pair—or else the same sentence-context pair but with the sentence negated—should appear in each sentence-context-world. Thus (we might hope) we can reasonably say that our sentence-context-worlds are complete.[17]

But what exactly is a context? We can point to particular contexts of utterance—such as the context that I am in right now as I make the utterances on this page—but we want to include not just actual contexts but possible contexts too. How can we specify all the possible contexts? We

---

[16] With thanks to Nicolas Cote for this visualization.

[17] Another way to put this is to say that each world should contain every possible sentence token—unless it contains its negation. Sentence tokens in this sense include all the information about the context of utterance. Thanks to Bjerring for this way of putting the point.

might try thinking of a context as a pair consisting of a time and place, and then we might list the possible contexts just by listing time-place pairs in every possible combination. But often the time and place in themselves are not enough to disambiguate an utterance. Consider, for example, the sentence 'the mountains there are much better than the ones we're used to'. To disambiguate this sentence, fixing the time and place of utterance is not enough. We also need to know which place is being talked about (the mountains *where*?); and we need to know who the 'we' refers to; and we need to know what standards are being assumed (much better in what sense?). In general 'there is no way of specifying a finite list of contextual co-ordinates' (Cresswell, 1972, p. 8).

For this and other reasons, I turn to an alternative view of sentence-worlds, on which the sentences that make up a world are not sentences in any natural language, but rather sentences in some artificial 'world-building' language.[18]

8.6 A world-building language

We saw in section 8.4 that we cannot replace metaphysically possible worlds with sentence-worlds, where a sentence-world is simply a set of natural language sentences, because natural language sentences can be ambiguous. In section 8.5 we tried to mend things by switching from sentence-worlds to sentence-context-worlds, where a sentence-context-world contains pairs each consisting of a natural language sentence together with a context. We saw some of the challenges that this account would face. Here I turn to an alternative: instead of natural language sentences, on this approach our sentence-worlds contain sentences in some artificial 'world-building' language. And we can ensure that the world-building language is free from ambiguity, so that we can avoid the complications associated with involving contexts.[19]

---

[18] Another reason sometimes given for switching to an artificial language is to ensure that we have the resources to generate enough worlds. Arguably there are more than continuum-many worlds—i.e. more worlds than there are rational numbers. But in any natural language the number of words is finite, and though the number of sentences in a natural language may be infinite, there will be at most continuum-many sentences. We can avoid any worries about the resources of the language by switching to certain sorts of artificial language (Lewis, 1986b, p. 143).

[19] We can also ensure that the language has enough words so that every possibility (of which there are more than continuum-many) can be represented.

182 STATES AS SOMETHING ELSE

One proposal for a world-building language is the 'Lagadonian' method, in which each object serves as the name of itself, and similarly each property or relation serves as a predicate representing itself (Lewis, 1986b). But this approach has faced a number of challenges. One is the problem of non-existent objects. To see the problem here, consider that it is possible that some object exists that does not actually exist. To take Jago's example, I could have had a sister (Jago, 2014, p. 129)—and so there should be possible worlds at which this sister exists and has various properties. Yet the Lagadonian language contains no name for this sister, for there is no actually existing object to serve as her name, so how can this possibility be represented? The same issue arises (arguably more severely) for non-existent properties. Theorists have put forward various responses, some of which run into difficulties over 'haecceitism'—a view on which two distinct possibilities can be qualitatively identical, the only distinction between them being that two objects or properties have swapped places (Melia, 2001; Sider, 2002; Jago, 2014). This issue raises many interesting questions, but here I focus on a different problem for the Lagadonian approach—a problem that is particularly relevant to this book.

The problem is that the Lagadonian language has only one name for each object. Thus, for example, there is only one name for George Orwell—that name being the man himself. For our purposes, this is a serious problem. The broad strategy of this chapter is to take the objects of credence to be sets of states, and to take those states to be more fine-grained than metaphysically possible worlds. The reason for taking states to be more fine-grained than metaphysically possible worlds is to accommodate the fact that the objects of credence are opaque: Tom can have a high credence that George Orwell is a writer, while having a low credence that Eric Blair is a writer, and yet the metaphysically possible worlds at which George Orwell is a writer are exactly the same as the metaphysically possible worlds at which Eric Blair is a writer. We turned to more fine-grained worlds in the hope that there might be fine-grained worlds where George Orwell is a writer and Eric Blair is not, or vice versa. But on the view we are currently exploring, a world is a set of sentences in the Lagadonian language, in which every object has just a single name which is itself. In this language it seems that there could not be a sentence stating that George Orwell is a writer that did not also state that Eric Blair is a writer, and so our hope for a world where George Orwell is a writer and Eric Blair is not (or vice versa) is dashed.

There are various ways that we could try to respond to this point. Mark Jago suggests that we might replace names with 'non-attributive

A WORLD-BUILDING LANGUAGE   183

property-bundles' (Jago, 2014, p. 158). This approach is related to Russell's descriptivist theory of names as discussed in chapter 2, but aims to avoid some of the objections to that view. An alternative move would be to increase the number of names in the Lagadonian language. Thus, for example, we might have two names, both referring to the same person: the first name is the pair consisting of George Orwell (the person himself) and the number 1; the second name is the pair consisting of George Orwell (again) and the number 2 (Jago, 2012, pp. 68–9). More generally, there will be numerous names for every object, each a pair consisting of the object together with an integer.[20] But how should these names be interpreted? What, for example, is the meaning of the name that consists of me and the number 3? The original Lagadonian language was easy to interpret, for each object and property represented itself. But this more complex language with multiple names for each object requires a full and detailed interpretation. We might think that this is merely an administrative task: surely all we need to do is list all the names and predicates in each natural language and then assign to each a pair consisting of the relevant object or property, along with an arbitrarily assigned number? But this will not do, because names and predicates in natural languages are ambiguous, picking out different objects and properties in different contexts. That was part of the motivation for switching to an artificial language—and to play its role the artificial language needs to be free from ambiguity. We cannot, then, give a full interpretation of our artificial language by simply matching its expressions up with ambiguous natural language expressions. Perhaps we could instead match expressions in our artificial language with pairs consisting of the relevant natural language expression together with a suitable context of utterance? But here we are losing the advantage that switching to the artificial language was supposed to bestow: we have avoided including contexts in our sentence-worlds, but now need to bring them back in to interpret our artificial language.

In short, there are challenges for this approach whereby worlds are sets of sentences in an artificial language. We also found in the previous section that there are challenges for an approach whereby worlds are sets of natural language sentence-context pairs. These challenges may not be decisive

---

[20] A similar move is made for properties. An agent might have a high credence that a cup contains water without having a high credence that it contains H_2O—even though (arguably) 'water' and 'H_2O' correspond to the same property. To accommodate this we might have multiple predicates in our world-building language, each consisting of the relevant property together with a number.

184   STATES AS SOMETHING ELSE

arguments against these approaches, but responding to these challenges is not easy. There are further questions to consider, however, so for the rest of this chapter I skate over this issue, writing as though worlds are simply sets of sentences in some language or other, and bracketing the problems with that approach. In the next section, I turn to the questions of completeness and coherence.

8.7 Completeness and coherence

In this section I return to a question that I raised briefly in section 8.4: what restrictions are there on the sentences that should be included in a sentence-world? Should we require completeness?[21] And should we require coherence? And what exactly would these requirements amount to?

In section 8.5, in which I considered sentence-worlds containing sets of natural language sentences, I said that on one way of spelling out completeness, we might require every sentence-world to contain every grammatical sentence in every language, unless it contains its negation. In section 8.6 I considered sentence-worlds consisting of sets of sentences in an artificial world-building language. Using this artificial language, we might construct our worlds using just atomic formulae and their negations. Thus we might require that for every n-place predicate, and for every combination of n names that will complete that predicate, each sentence-world will contain either the resulting atomic formula or the negation of that atomic formula.[22] Rudolf Carnap constructs his worlds in this way, using only atomic formulae and their negations (Carnap, 1947). If we wished, we might also require that our worlds contain every sentence (or else its negation) that can be constructed from atomic formulae together with a stock of variables, the quantifiers ($\forall$, $\exists$), and the logical connectives ($\neg$, $\vee$, $\wedge$, $\rightarrow$, $\leftrightarrow$). In this way effectively we would require that a sentence-world contains every grammatical sentence in this artificial language, unless it contains that sentence's negation.

---

[21] I note here that some theorists—such as (Barwise and Perry, 1983)—consider incomplete 'situations' to be superior for many purposes to complete worlds, where an incomplete situation need not include every atomic formula or its negation. I do not explore this interesting literature further here.

[22] Specifying what counts as the negation of a sentence is simple enough in an artificial language that contains '$\neg$', but more of a challenge in natural languages like English. Which sentences count as the negation of 'grass is green'? Plausibly 'It's not the case that grass is green' does, but what about 'grass is not green', or 'no way is grass green!'?

COMPLETENESS AND COHERENCE   185

The issue of completeness is relevant to an issue that I touched on in section 8.4: how do belief and credence attribution statements work on this approach? Take, for example, the following belief attribution statement:

(8g)   Tom believes that grass is green and the sky is blue.

This states that Tom stands in the belief relation to a particular object, and on the view that we are currently exploring that object is a set of sentence-worlds. Which sentence-worlds? Are they simply the sentence-worlds that contain the sentence 'grass is green and the sky is blue'? Or is it all the sentence-worlds whose sentences *imply* the sentence 'grass is green and the sky is blue'? A theorist's views on this will depend on what she takes sentence-worlds to be like. If she takes sentence-worlds to contain every grammatical sentence in every language unless it contains that sentence's negation, then she can say that (8g) states that Tom stands in the belief relation to the set of sentence-worlds that contain the sentence 'grass is green and the sky is blue'. But what if she takes sentence-worlds to contain many fewer sentences? Suppose, for example, that she takes sentence-worlds to contain just atomic sentences in some specific artificial language? Then there will be no sentence-worlds that contain the sentence 'grass is green and the sky is blue', and so (8g) will be pointing us towards an object of belief that is the empty set. Indeed, all belief and credence attribution statements written in any natural language will be attributing belief or credence in the empty set. Thus a theorist who holds that sentence-worlds contain just sentences written in some artificial language—or indeed any theorist who holds that a sentence-world need not include every grammatical sentence in every language unless it contains that sentence's negation—will need to say that belief and credence attribution statements work by pointing us towards the set of worlds that *imply* the relevant sentence without necessarily containing it. And we will need to understand implication in such a way that it encompasses translation—so that sentences in an artificial language can imply their translations into any natural language. Thus, for example, (8g) states that Tom stands in the belief relation to the set of sentence-worlds, such that for each such sentence-world, the sentences contained in it jointly imply the sentence 'grass is green and the sky is blue'. Of course this idea of 'implying' needs to be spelt out—and I return to this idea shortly.

Besides the question of whether and in what sense a sentence-world ought to be complete, we can also consider whether and in what sense a sentence-world ought to be coherent. We might require, as a first rule, that no world

186 STATES AS SOMETHING ELSE

should contain both a sentence and its negation. Graham Priest rejects this rule and defends a very liberal account on which there are worlds that contain both a sentence and its negation (Priest, 2005). Priest has reasons for taking this view seriously, but for users of the credence framework there would be very surprising implications of taking states to be worlds in this sense. For example, recall that in chapter 3 I showed that the probability axioms entailed that for any rational credence function Cr and any proposition P, $Cr(\neg P) = 1 - Cr(P)$. The argument for this relied on the assumption that the set of states where $\neg P$ holds is the complement of the set of states where P holds. If we take states to be worlds in Priest's sense, then this assumption would be incorrect, for there are worlds where both P and $\neg P$ hold.[23] In general, if we take a very liberal attitude towards worlds, then the restrictions imposed by the probability axioms will look very different from (and much weaker than) the sorts of restrictions that users of the credence framework would expect. For users of the credence framework, then, it makes sense to rule that no sentence-world should contain both a sentence and its negation.

Should any more than this be required? Here the issues of completeness and coherence interact. We might require—as a step towards completeness—that a world should be closed under natural deduction.[24] That is, if a world contains a set of sentences (the 'premises') from which a sentence (the 'conclusion') follows by natural deduction, then this sentence (the 'conclusion') should be included in the world. And, of course, if this sentence (the 'conclusion') is included, then the negation of the conclusion must be excluded (by our first rule, that no world should contain both a sentence and its negation).

Many theorists working in this area stop short of requiring this sort of closure under natural deduction. In some cases this is because they are trying to solve the problem of logical omniscience. To illustrate the problem that they are trying to solve here, we can consider an agent who has a high credence that P. P is logically equivalent to the claim that $(\neg P \rightarrow (P \wedge \neg P))$, though the agent (not being logically omniscient) has not worked this out and has a low credence that $(\neg P \rightarrow (P \wedge \neg P))$. On the view that we're

---

[23] Priest also rejects 'completeness', and so as well as worlds where both P and $\neg P$ hold, there will also be worlds where neither P nor $\neg P$ hold.

[24] What follows by natural deduction depends on the details of the system that you are working with, and a variety of natural deduction systems have been proposed with no consensus that any particular such system is the one true system. Furthermore, there are alternatives to natural deduction which might be substituted here.

considering, the objects of credence are sets of worlds. So our agent has a high credence in the set of worlds that contain P, and a low credence in the set of worlds that contain $(\neg P \rightarrow (P \land \neg P))$. But if worlds are closed under natural deduction, then the set of worlds that contain P is identical to the set of worlds that contain $(\neg P \rightarrow (P \land \neg P))$, and so we seem to be forced to say that the agent has both a high credence and a low credence in the very same object.

To handle this problem (and for further reasons) theorists have developed accounts on which sentence-worlds are required to have a certain sort of minimal coherence, without being required to be closed under natural deduction. On this view, a world will not contain directly contradictory sentences—i.e. a sentence and its negation—and in addition, worlds are closed under *obvious* logical consequence; but worlds are not closed under *non-obvious* logical consequence. If this could be made to work, then it might be thought to solve the problem of logical omniscience, while still giving the probability axioms some bite. However, as Jens Christian Bjerring and Wolfgang Schwarz show (Bjerring, 2013; Bjerring and Schwarz, 2017), there is a serious problem with this approach. To see the problem, take the case above of an agent who has a high credence that P but a low credence that $(\neg P \rightarrow (P \land \neg P))$. Though P does logically entail $(\neg P \rightarrow (P \land \neg P))$, the entailment is not obvious,[25] and so it seems that an agent who is not logically omniscient might reasonably have a higher credence in P than in the complex claim that it entails. It seems that we can accommodate this by having P true at some worlds where $(\neg P \rightarrow (P \land \neg P))$ is false. The problem with this idea is that—because $(\neg P \rightarrow (P \land \neg P))$ follows from P by natural deduction—there is a proof with P as the only premise and $(\neg P \rightarrow (P \land \neg P))$ as the conclusion, where each step of the proof is obvious. Though entailment is transitive (if P entails that Q and Q entails that R, then P entails that R), *obvious entailment* is not transitive, and a proof can lead from a premise to a conclusion that does not obviously follow via a series of obvious steps. Given that worlds are closed under obvious logical entailment, at any world where the premise is true, the first sentence derived in the proof must also be true there as it follows obviously from the premise; and the second sentence derived in the proof—which follows obviously from the premise together with the first sentence derived—must therefore also be true at that world; and so on—with the result that the conclusion must also be true there. Thus,

---

[25] Of course, if this entailment seems too obvious, a more complex claim can be used to make the same point.

188 STATES AS SOMETHING ELSE

given that worlds are closed under *obvious* logical entailment, it follows that at any world where P is true, each sentence of the proof must be true—including the conclusion ($\neg P \rightarrow (P \wedge \neg P)$). It seems, then, that every world where P is true is also a world where ($\neg P \rightarrow (P \wedge \neg P)$) is true, and so if the objects of credence are sets of worlds, and worlds are closed under obvious logical entailment, then if an agent has a high credence in P then her credence in ($\neg P \rightarrow (P \wedge \neg P)$) must also be high.

These are serious problems for theorists trying to solve the problem of logical omniscience. In this book, though, I am not trying to solve the problem of logical omniscience, but the problem of opacity. This problem seems to me to be much more urgent. We might more or less plausibly handle the problem of logical omniscience by saying that the credence framework is designed to represent the epistemic states of *ideally* rational agents, but we cannot plausibly handle the problem of opacity in this way. Even an ideally rational agent is not required to know every identity claim: to know every such claim would be to have something close to ordinary (rather than logical) omniscience, and the credence framework has no relevance for omniscient agents. For our purposes, then, given that we are not trying to solve the problem of logical omniscience, we can straightforwardly require that sentence-worlds be closed under natural deduction. A more pressing question for us is whether natural deduction is enough, and I turn to this question in the next section.

8.8 A stronger notion of coherence

In chapter 2 I discussed and dismissed the idea that the objects of belief and credence might simply be sentences.[26] One of the reasons I gave for rejecting this claim was that the following three belief attribution claims are either translations (that is, they mean the same thing), or harmless rephrasings of each other:

(8h) Tom believes that the apple is in the bowl.

(8i) Tom believes that the bowl has the apple in it.

(8j) Tom croit que le pomme est dans le bol.

---

[26] For similar reasons, I do not explore an alternative version of the credence framework on which the objects of credence are simply sentences, and the relevant algebra is a Boolean algebra.

A STRONGER NOTION OF COHERENCE   189

If the objects of belief are sentences, then it is unclear why (8h)–(8j) are translations or harmless rephrasings of each other. (8h) presumably states that Tom stands in the belief relation to the sentence 'the apple is in the bowl'. Why would it follow that he also stands in the belief relation to the sentence 'the bowl has the apple in it'? These are two different sentences, after all. The answer, of course, is that though these are indeed two different sentences, they mean the same thing—that is, they have the same content. We are thus naturally led to the view that the objects of belief are not sentences, but rather the content of sentences (as uttered in context)—i.e. propositions. Any statement attributing a belief to an agent can be harmlessly rephrased by replacing the sentence following the 'that' clause with an alternative sentence that has the same content (in context).[27]

The implications of this point for the view we are considering in this chapter depend on exactly how that view is spelt out. Let us begin by considering the view on which each sentence-world contains every grammatical sentence in every language—unless it contains that sentence's negation. In that case, to say 'Tom has a high credence that the apple is in the bowl' is to state that Tom stands in a certain relation to an object of credence, and that object of credence is the set of sentence-worlds which contain the sentence 'the apple is in the bowl'. Similarly, to say 'Tom has a high credence that the bowl has the apple in it' is to state that Tom stands in a certain relation to the set of sentence-worlds which contain the sentence 'the bowl has the apple in it'. Given that our two statements about Tom seem to be harmless rephrasings of each other, it follows that both credence attribution statements must be pointing to the same object of credence. In other words, the set of sentence-worlds containing 'the apple is in the bowl' must be the same as the set of sentence-worlds containing 'the bowl has the apple in it'. More generally, each sentence-world must be closed under sameness of content: if a sentence-world has a sentence in it that has the same content as some other sentence, then that other sentence must also appear in that sentence-world.[28]

---

[27] Bjerring and Schwarz give a persuasive example to similarly show that the objects of *discovery* are not sentences: '...consider Euclid's discovery that there are infinitely many primes. Presumably, the content of Euclid's discovery can be expressed not only by "there are infinitely many primes" but also by trivially equivalent statements such as "the number of primes is infinite": that the number of primes is infinite is not a further discovery, also made by Euclid' (Bjerring and Schwarz, 2017, p. 30).

[28] See also the discussion of 'cognitive equivalence' in (Bjerring and Skipper, 2020).

190   STATES AS SOMETHING ELSE

But even being closed under both sameness of content and closed under natural deduction is not enough. If Tom has a high credence that the apple is in the bowl, then he will also have a credence at least as high that the apple is not beside the bowl. It seems, then, that any sentence-world containing 'the apple is in the bowl' must also contain 'the apple is not beside the bowl'. But how is this to be guaranteed? It is not as though 'the apple is in the bowl' and 'the apple is not beside the bowl' have the same content, and there is no formal deductive proof from 'the apple is in the bowl' to 'the apple is not beside the bowl'. It seems that we need the more general rule that sentence-worlds should be closed under implication—where implication goes well beyond natural deduction. We might achieve this by requiring each world to contain a certain set of axioms (Lewis, 1986b, p. 153).[29] The set might include, for example, the axiom that for any pair of objects x and y, if x is in y then x is not beside y—and numerous other axioms would also need to be included. Then, as each world contains this set of axioms, and as each world is closed under natural deduction, it follows that each world is closed under implication (and under sameness of content) in the relevant way. Of course the challenge would be to specify the relevant set of axioms.[30] I do not attempt to meet this challenge here, but just note a couple of directions that should be avoided. It would not work to claim that the axioms are all and only those sentences that express metaphysical necessities—for obviously we do not want the axioms to include sentences such as 'George Orwell is Eric Blair', for then every world would contain this sentence and we would be unable to make sense of an agent who had a high credence that George Orwell is a writer and a low credence that Eric Blair is a writer. And obviously it would not work to claim that the axioms are all and only those sentences whose truth follows given natural deduction. One natural thought is that the answer here is to say that a sentence is an axiom iff it should be assigned a credence of 1 by any rational agent. I return to this natural thought in the next section.

Here I consider the alternative view on which worlds need not contain, for every grammatical sentence, either that sentence or its negation. On this view, for example, worlds might contain only sentences in some artificial

---

[29] These axioms bear a close relation to Carnap's 'meaning postulates' (Carnap, 1952), and have connections to the more recently discussed 'laws of metaphysics' (Kment, 2014; Schaffer, 2017).

[30] There are reasons to think that the choice of axioms ought to depend on context, and in particular on the agent to whom beliefs or credences are being attributed, and Bjerring and Schwarz show that this creates additional challenges for this view (Bjerring and Schwarz, 2017).

language, and perhaps even just the atomic sentences—or even just some small subset of the atomic sentences—of that language. These theorists will need to understand credence and belief attribution statements somewhat differently. For example, statement (8h) above—'Tom believes that the apple is in the bowl'—states that Tom stands in the belief relation to a certain object of belief, and that object of belief is a set of sentence-worlds. The relevant set of sentence-worlds cannot be those that contain the sentence 'the apple is in the bowl', for on the view we are considering, that would simply be the empty set. Rather, the relevant set of sentence-worlds must be those whose sentences *imply* the sentence 'the apple is in the bowl'. But what exactly is implication? Natural deduction alone will not be enough: there is no deduction from a sentence in an artificial language to the English sentence 'the apple is in the bowl'. We might introduce a set of axioms, and state that a set of sentences implies some given sentence iff that sentence can be logically deduced from that set of sentences together with the axioms. Evidently here we face a problem specifying what is involved in implication that parallels the problem faced in the previous paragraph with specifying what coherence requires. As Lewis writes: 'It matters little whether we take a rich language and face a problem of saying which sets of its sentences are consistent, or whether instead we take a poor language and face the problem of saying what its sets of sentences imply' (Lewis, 1986b, p. 152). Either way, then, this is a challenge for theorists who adopt the strategy described in this chapter. And on the grounds that it makes little difference, for the rest of this chapter I assume the second view, on which a sentence-world contains just every atomic sentence in a 'world-building' language, or its negation.

8.9 Arguments for probabilism

I turn now to consider the probability axioms, and the arguments for those axioms that I set out in chapter 3. I stated there that the interpretation of the axioms depended on the general interpretation of the probability framework. As an illustration, consider axiom 2 (Kolmogorov, 1933 (1950)): the probability of Ω is 1.

As I've been doing throughout this book, I'll interpret the probability framework as a credence framework, and here I'll consider what this axiom means given the approach we are following in this chapter. In this chapter, we are interpreting states as sentence-worlds, where each sentence-world is a

192 STATES AS SOMETHING ELSE

set of sentences. Let's consider, then, the following credence attribution claim:

(8k)   Tom has a credence of 0.9 that a bachelor is an unmarried man.

Given (8k), does Tom's credence function violate axiom 2? Well, that depends. (8k) tells us that Tom has a credence of 0.9 in some object of credence, and that object of credence is a set of sentence-worlds. Which set of sentence-worlds? The relevant set of sentence-worlds will be the set of those sentence-worlds such that the sentences in that world imply the sentence 'a bachelor is an unmarried man'. Is this *all* the sentence-worlds? That is, is this Ω? If so, then Tom's credence function does indeed violate axiom 2. But do all sentence-worlds imply 'a bachelor is an unmarried man'? That will depend on whether that sentence (or some closely related sentence) is included as an axiom. As we saw in the last section, there is no obvious way to settle which sentences should be included as axioms—though perhaps the most natural approach is to include as axioms all sentences such that intuitively any rational agent who understood the sentence would assign it a credence of 1.

We can see, then, that decisions about what sorts of credence functions are rational feeds into decisions about the nature of states, and as a result the arguments for probabilism lose their bite. Consider, for example, the dutch-book argument, as described in chapter 3. The underlying idea is that only an irrational agent would accept as fair a set of bets that is guaranteed to result in a loss. And in order to get a dutch-book argument for the probability axioms, 'guaranteed' needs to mean 'at every state'. For the argument to work, we ought to have an independent account of the nature of states, feel intuitively compelled to accept that no rational agent would accept a set of bets that loses money at every state, and thus be forced to accept that an agent who violates the probability axioms is irrational. But in fact our judgements about rationality help mould our interpretation of states in the first place, and the conclusion of a dutch-book argument will simply reflect these judgements. To see this, suppose that you are presented with a dutch-book argument, the conclusion of which is that a rational agent must have a credence of 1 that a bachelor is an unmarried man. And suppose that you disagree with this conclusion: you can think of people who are perfectly rational, and yet are less than perfectly certain of this claim. In that case, you can just adjust your interpretation of states, denying that the sentence 'a bachelor is an unmarried man' (or any closely related sentence) is an axiom.

STRUCTURED PROPOSITIONS   193

Given that adjustment, the agent in this scenario is not guaranteed to lose money, for there are states that do not imply the sentence 'a bachelor is an unmarried man'. The argument thus has no force: you reach the conclusion that you antecedently accepted by adjusting your interpretation of states accordingly. A similar point holds for both the accuracy argument and the representation theorem (Mahtani, 2020). On this approach, then, the dutch-book argument, accuracy argument, and representation theorem have no force as arguments for probabilism. Motivating the probability axioms is thus a further challenge for this approach.

8.10 Structured propositions

I finish this chapter with a brief discussion of an alternative approach. Amongst those who think that sets of metaphysically possible worlds are too coarse to serve as the objects of belief/credence, we can distinguish two sorts of theorists: those who think that the objects of belief/credence are sets of some sort of more fine-grained worlds; and those who think that the objects of belief/credence are structured propositions. So far in this chapter I have focused on the first sort of view, but in this section I turn to the second.

I introduced the idea of structured propositions in chapter 2. The general idea is that a proposition is a collection of objects and properties structured in a particular way. Typically the structure of a proposition parallels the structure of the sentence that expresses it. Here is an example to illustrate the view. We can take the following two sentences: 'Paris is big and Lincoln is small' and 'Lincoln is small and Paris is big'. These two sentences express propositions that each contain the same objects (Paris, Lincoln) and the same properties (bigness, smallness), but they are differently structured. Thus the sentences express different propositions. We can see here a contrast with the view on which propositions are sets of metaphysically possible worlds. The metaphysically possible worlds where Paris is big and Lincoln is small are the same as the metaphysically possible worlds where Lincoln is small and Paris is big, so on this view they express the same proposition.

One of the advantages of a structured proposition account is that it has at least a partial answer to the problem of logical omniscience. Take our agent who has a high credence in some simple claim, P, but a low credence in a complex but logically equivalent claim ($\neg P \rightarrow (P \wedge \neg P)$). The simple claim and the complex but logically equivalent claim are true at the same metaphysically

194 STATES AS SOMETHING ELSE

possible worlds—and so if we take the objects of credence to be sets of metaphysically possible worlds, then we seem forced to say that an agent's credence in P must be the same as her credence in $(\neg P \to (P \wedge \neg P))$.[31] And this strikes many as counterintuitive. On the structured proposition account, we have a ready answer: 'P' and '$(\neg P \to (P \wedge \neg P))$' express different propositions with quite different structures. Thus we can make sense of the idea of an agent having a high credence in one and a low credence in the other.

For our purposes, however, there are two immediate problems with the idea of structured propositions. The first is that it is hard to see how to fit them into the credence framework. On the credence framework, the objects of credence are sets of states, but structured propositions do not seem to be sets of anything. This is a problem that I return to shortly. The other problem is that though structured propositions seem to help with the problem of logical omniscience, they do not help with the problem of opacity. To see this, consider that the following two sentences express identical propositions on the structuralist's account, for the propositions are composed of the same objects in the same structure: 'George Orwell is a writer'; 'Eric Blair is a writer'. Thus it is mysterious how Tom might have a high credence in one but a low credence in the other. Structuralists are aware of this problem of course, and have various things to say in response (for example, along the lines of Salmon's account discussed in chapter 4)—but it seems clear that taking propositions to be structured entities does not in itself help at all with this problem (Ripley, 2012). For our purposes, then, it seems that there is no need to switch to a view on which propositions are structured, and so we can continue to claim—as is congenial to users of the credence framework—that propositions are sets of states.

But there is a potential problem for this broad strategy. Soames has an argument designed to show that propositions simply cannot be sets of states: they cannot be sets of metaphysically possible worlds, for reasons that are familiar, but Soames aims to establish quite generally that propositions do not correspond to sets of states (or 'circumstances')—even if those states are taken to be more fine-grained than metaphysically possible worlds (Soames, 1987). Soames puts forward various versions of his argument, relying on different assumptions. One such version relies on the assumption that

---

[31] Chalmers's distinction between primary and secondary intensions is less immediately helpful here than it seemed to be in our standard opacity cases, because the set of (centred) worlds that verify P are the same as the set of (centred) worlds that verify $(\neg P \to (P \wedge \neg P))$. Chalmers's own approach to this sort of problem is to claim that the objects of credence are enriched propositions, which are structured objects. See section 7.8.

names are directly referential. Given this assumption, we can move, for example, from the claim that Tom believes that 'George Orwell' refers to George Orwell, to the claim that Tom believes that 'George Orwell' refers to Eric Blair. If this is combined with the assumption that the objects of belief are sets of states, and given a range of other natural assumptions about states (such as, for example, the claim that if P and Q both hold at a state, then the conjunction also holds), we arrive at the implausible conclusion that Tom believes that 'George Orwell' and 'Eric Blair' are co-referential. Soames locates the error here in the assumption that the objects of belief are sets of states. It is open to us to instead locate the error in the assumption that names are directly referential: Soames has some powerful arguments in favour of this assumption, but we have seen that for our purposes there are also powerful reasons to think that names—at least as they figure in credence attribution statements—are not directly referential. Soames gives several other versions of his argument: for example, a different version can be constructed assuming that demonstratives rather than names are directly referential. Again—for our purposes we also have reasons to reject this claim as far as credence attribution claims are concerned. There are options, then, for rejecting Soames's argument (and see (Ripley, 2012; Elbourne, 2010) for detailed discussions).

But before leaving this issue, I turn to the following question: what if Soames is right, and the objects of belief and credence are indeed structured propositions? At first that view seems to be incompatible with the credence framework. For on the credence framework, the objects of credence are taken to be sets of states, and sets of states do not seem to be structured in the way that Soames requires. Nevertheless, there is a way to accept Soames's view that propositions are structured while also maintaining that the objects of credence are sets of states. We just need to radically alter our view of what constitutes a state. Suppose, for example, that a state contains the sentence 'Paris is big and Lincoln is small'. It is natural to assume that this state must also contain the sentence 'Lincoln is small and Paris is big', and indeed this follows if we assume that states are closed under natural deduction. But we might drop the requirement that states should be closed under natural deduction and instead require merely that states should be closed under identity of proposition expressed. In that case, a state containing the sentence 'Paris is big and Lincoln is small' need not also contain the sentence 'Lincoln is small and Paris is big': one follows logically from the other, but the propositions expressed have different structures. On the other hand, a state that contains the sentence 'George Orwell is a writer' will also contain

196 STATES AS SOMETHING ELSE

the sentence 'Eric Blair is a writer', on the grounds that the two sentences express propositions with the same structure: the propositions expressed contain the same objects and properties structured in the same way. Of course, this is not a view that I favour: I claim that an agent's credence that George Orwell is a writer can differ from her credence that Eric Blair is a writer—a view that cannot be accommodated on this picture. The point here is just that given a sufficiently liberal view of states, it may be possible for the credence framework to be interpreted in a way that is congenial to those who take the objects of credence to be structured propositions.[32]

8.11 Chapter summary

In this part of the book, I've been considering how the credence framework can be interpreted so that it can handle our tenet that credence claims are opaque. In this chapter I've focused on the idea of interpreting states as something other than metaphysically possible worlds. One natural proposal along these lines is to take states to be sets of sentences. On this proposal, there are many details to be filled in and challenges to be met. What language or languages are the sentences in? How does context fit into this picture? Should the worlds be complete and/or coherent, and in what sense? How these questions are answered will have numerous repercussions for users of the credence framework. For example, whether two objects of credence are independent, or whether they are disjoint, or whether an object of credence is a tautology (Ω) will all depend on how these questions are answered. And I have argued that—on a natural way of spelling out some of these details—the arguments for probabilism lose their force. Thus, though this may be a promising way of interpreting the credence framework so that it can handle our tenet, users of the framework should be alert to the wide-reaching implications of this approach.[33]

---

[32] It is an open question whether the credence framework would also be able to accommodate still other views in the literature, such as 'hidden indexical' theories (Crimmins and Perry, 1989; Schiffer, 1995) and contextualist theories (Richard, 1990).

[33] A further implication, which I do not explore here, concerns Leibniz's Law (Leibniz, 1666–1716 (1969)). Leibniz's Law states that where x and y are identical, if x has any property then y must have that property too. It follows from Leibniz's law, together with the claim that George Orwell is identical to Eric Blair, that George Orwell and Eric Blair share the same properties—including arguably the property of being believed by Tom to be in the café. Some of the logical implications of rejecting Leibniz's Law are explored in (Bacon and Russell, 2019; Caie, Goodman, and Lederman, 2020).

9

Conclusion

What should a user of the credence framework take away from this book?

Firstly, credence claims are opaque. To accept this is to accept that a rational agent can have a high credence that George Orwell is a writer and a low credence that Eric Blair is a writer—and so on for a multitude of similar cases. When we consider the credences that an agent has about some person or object, it matters how that person or object is designated. I argued for this tenet in chapter 4.

Secondly, this simple tenet has wide-ranging and important implications for users of the credence framework. In chapters 5 and 6 I showed that there are implications for the reflection and deference principles, for the Principal Principle, for decision theory, and for various views in welfare economics. These are some examples of the implications, and doubtless there are many other implications to be explored.

Thirdly, there is no easy way for the underlying framework to accommodate the tenet that credence claims are opaque. In chapters 7 and 8 I tried out several approaches. Some showed potential, but all had ramifications: in particular, a promising approach based on two-dimensionalism had implications for deference principles and conditionalization, and an approach involving impossible worlds had implications for arguments for probabilism.

An important aim of this book is to call attention to these points. By no means do I think that the credence framework should be abandoned. The credence framework is a powerful, flexible tool. It is used by a vast number of theorists working across a wide range of different disciplines. It is entirely standard to use it to inform policy choice and in many other practical situations, and there is no obvious alternative framework with the same sort of power and reach that could be used in its stead.

My hope is rather that the credence framework will continue to be used—but with an awareness of the issues that I have raised in this book. When a formal epistemologist sets out an agent's credence function, she should bear in mind that the credences assigned may depend on how the relevant objects are designated. She should ask herself: does anything rest on how I have chosen to designate these objects? Would the credences assigned be different

The Objects of Credence. Anna Mahtani, Oxford University Press. © Anna Mahtani 2024.
DOI: 10.1093/oso/9780198847892.003.0009

198 CONCLUSION

if I designated these objects differently? And can I justify my choice of designators? Similarly, when a decision theorist sets out a decision table, she should ask herself how she is designating the objects (and predicates) in giving the states, actions, and outcomes. Does anything ride on this? Can the choice be defended? And for welfare economists, a key question is: when considering the wellbeing of a range of people, how am I designating them, and why? And all theorists should have in mind that the fundamental interpretation of the credence framework is as yet unsettled: it is not obvious how it can be interpreted to accommodate the central claim of this book. Should a theorist's view depend on some particular interpretation, then the implications of this should be carefully considered.

Here I am swimming against the tide. Recently, philosophers have been encouraged to take their work outside academia and see how it can be applied to practical problems. In contrast, I am urging users of the credence framework—who are typically already deeply concerned with practical problems—to be aware of an issue that has so far been of predominant interest to philosophers of language. But I see this as an example of how philosophy can be useful: an insight—recognized and discussed since Frege—has implications that reach out, through the credence framework, to choices that are made with vast practical repercussions. All who use the framework should be aware of the insight and have a sense of its ramifications.

References

Adams, R. M. (1974). Theories of Actuality. *Nous*, 8, 211–31.

Adler, M. (2017). *Measuring Social Welfare: An Introduction*. Manuscript.

Adler, M., and Holtug, N. (2019). Prioritarianism: A Response to Critics. *Politics, Philosophy and Economics*, 18 (2), 101–44.

Arntzenius, F. (2003). Some Problems for Conditionalization and Reflection. *Journal of Philosophy*, 100 (7), 356–70.

Bacon, A., and Russell, J. (2019). The Logic of Opacity. *Philosophy and Phenomenological Research*, 99 (1), 81–114.

Barwise, J., and Perry, J. (1983). *Situations and Attitudes*. Cambridge, MA: MIT Press.

Beaney, M. (1996). *Frege: Making Sense*. London: Gerald Duckworth & Co.

Beddor, B. (2020). Certainty in Action. *The Philosophical Quarterly*, 70 (281), 711–37.

Berto, F., and Jago, M. (2019). *Impossible Worlds*. Oxford: Oxford University Press.

Bigelow, J., Collins, J., and Pargetter, R. (1993). The Big Bad Bug: What Are the Humean's Chances? *The British Journal for the Philosophy of Science*, 44 (3), 443–62.

Bjerring, J. (2013). Impossible Worlds and Logical Omniscience: An Impossibility Result. *Synthese*, 190 (13), 2505–24.

Bjerring, J., and Schwarz, W. (2017). Granularity Problems. *The Philosophical Quarterly*, 67 (266), 22–37.

Bjerring, J., and Skipper, M. (2020). Hyperintensional Semantics: A Fregean Approach. *Synthese*, 197 (8), 3535–58.

Bovens, L., and Rabinowicz, W. (2010). The Puzzle of the Hats. *Synthese*, 172 (1), 57–78.

Bradley, R. (2017). *Decision Theory with a Human Face*. Cambridge: Cambridge University Press.

Braun, D. (2002). Cognitive Significance, Attitude Ascriptions, and Ways of Believing Propositions. *Philosophical Studies*, 108 (1/2), 65–81.

Braun, D. (2016). The Objects of Belief and Credence. *Mind*, 125 (498), 469–97.

Briggs, R. (2009a). The Anatomy of the Big Bad Bug. *Nous*, 43 (3), 428–49.

Briggs, R. (2009b). Distorted Reflection. *Philosophical Review*, 118 (1), 59–85.

Broome, J. (1984). Uncertainty and Fairness. *The Economic Journal*, 94 (375), 624–32.

Broome, J. (1995). The Two-Envelope Paradox. *Analysis*, 55 (1), 6–11.

Brown, P. M. (1976). Conditionalization and Expected Utility. *Philosophy of Science*, 43 (3), 415–19.

Buchak, L. (2013). *Risk and Rationality*. Oxford: Oxford University Press.

200 REFERENCES

Caie, M., Goodman, J., and Lederman, H. (2020). Classical Opacity. *Philosophy and Phenomenological Research*, 101 (3), 524–66.

Cargile, J. (1992). On a Problem about Probability and Decision. *Analysis*, 52, 211–16.

Carnap, R. (1947). *Meaning and Necessity*. Chicago: University of Chicago Press.

Carnap, R. (1950). *Logical Foundations of Probability*. Chicago: University of Chicago Press.

Carnap, R. (1952). Meaning Postulates. *Philosophical Studies*, 3 (5), 65–73.

Chalmers, D. (2002). The St. Petersburg Two-Envelope Paradox. *Analysis*, 62 (2), 155–7.

Chalmers, D. (2004). Epistemic Two-Dimensional Semantics. *Philosophical Studies*, 118 (1/2), 153–226.

Chalmers, D. (2011a). Frege's Puzzle and the Objects of Credence. *Mind*, 120 (479), 587–635.

Chalmers, D. (2011b). The Nature of Epistemic Space. In A. Egan and B. Weatherson (eds), *Epistemic Modality* (pp. 60–107). Oxford: Oxford University Press.

Chalmers, D. (2011c). Propositions and Attitude Ascriptions: A Fregean Account. *Nous*, 45 (4), 595–639.

Christensen, D. (1991). Clever Bookies and Coherent Beliefs. *The Philosophical Review*, 100, 229–4.

Christensen, D. (1996). Dutch-Book Arguments Depragmatized: Epistemic Consistency for Partial Believers. *Journal of Philosophy*, 93, 450–79.

Cirilo de Melo, W., and Cussens, J. (2004). Leibniz on Estimating the Uncertain: An English Translation of De incerti aestimatione with Commentary. *The Leibniz Review*, 14.

Clark, M., and Shackel, N. (2000). The Two-Envelope Paradox. *Mind*, 109 (435), 415–42.

Clarke, R. (2013). Belief Is Credence One (in Context). *Philosophers' Imprint*, 13, 1–18.

Crane, T., and Mellor, D. H. (1990). There Is No Question of Physicalism. *Mind*, 99 (394), 185–206.

Cresswell, M. (1972). The World Is Everything That Is the Case. *Australasian Journal of Philosophy*, 50 (1), 1–13.

Crimmins, M., and Perry, J. (1989). The Prince and the Phone Booth. *The Journal of Philosophy*, 86, 685–711.

Davidson, D. (1963). Actions, Reasons and Causes. *The Journal of Philosophy*, 60 (23), 685–700.

Davidson, D. (1967). Causal Relations. *The Journal of Philosophy*, 64 (21), 691–703.

Davidson, D. (1973). Radical Interpretation. *Dialectica*, 27, 314–28.

de Finetti, B. (1931). Probabilism. English translation in *Erkenntnis*, 1989, 31, 169–223.

Diamond, P. A. (1967). Cardinal Welfare, Individualistic Ethics, and Interpersonal Comparison of Utility. *Journal of Political Economy*, 75 (5), 756–66.

Dietrich, F., and List, C. (2005). The Two-Envelope Paradox: An Axiomatic Approch. *Mind*, 114 (454), 239–48.

REFERENCES 201

Dorr, C. (2003). Sleeping Beauty: In Defence of Elga. *Analysis, 62* (276), 292–6.

Edgington, D. (2004). The Inaugural Address: Two Kinds of Possibility. *Proceedings of the Aristotelian Society, Supplementary Volumes, 78*, 1–22.

Elbourne, P. (2010). Why Propositions Might Be Sets of Truth-Supporting Circumstances. *Journal of Philosophical Logic, 39* (1), 101–11.

Elga, A. (2000). Self-Locating Belief and the Sleeping Beauty Problem. *Analysis, 60* (2), 143–7.

Elga, A., and Rayo, A. (Forthcoming). Fragmentation and Logical Omniscience. *Nous.*

Eriksson, L., and Hájek, A. (2007). What Are Degrees of Belief? *Studia Logica, 86*, 183–213.

Feldman, R. (2006). Epistemological Puzzles about Disagreement. In S. Hetherington (ed.), *Epistemic Futures* (pp. 216–36). New York: Oxford University Press.

Field, H. (1978). Mental Representation. *Erkenntnis, 13* (1), 9–61.

Field, H. (2001). *Truth and the Absence of Fact.* Oxford: Oxford University Press.

Fine, K. (2021). Constructing the Impossible. In L. Walters and J. Hawthorne (eds), *Conditionals, Paradox, and Probability: Themes from the Philosophy of Dorothy Edgington* (pp. 141–63). Oxford: Oxford University Press.

Fleurbaey, M. (2008). *Fairness, Responsibility, and Welfare.* Oxford: Oxford University Press.

Fleurbaey, M. (2010). Risky Social Situations. *Journal of Political Economy, 118* (4), 649–80.

Fleurbaey, M., and Voorhoeve, A. (2013). Decide as You Would with Full Information! An Argument against Ex Ante Pareto. In O. Norheim, S. Hurst, N. Eyal, and D. Wikler (eds), *Inequalities in Health: Concepts, Measures, and Ethics* (pp. 113–128). Oxford: Oxford University Press.

Fodor, J. (1987). *Psychosemantics.* Cambridge, MA: MIT Press.

Forbes, G. (1985). *The Metaphysics of Modality.* Oxford: Oxford University Press.

Frege, G. (1980 (1879–1903)). *Translations from the Philosophical Writings of Gottlob Frege.* (P. Geach and M. Black, trans.) Oxford: Blackwell.

Gallow, D. (2021). *Two-Dimensional Deference.* Manuscript.

Gillies, D. (2000). *Philosophical Theories of Probability.* Oxford: Routledge.

Greaves, H., and Wallace, D. (2006). Justifying Conditionalization: Conditionalization Maximizes Expected Epistemic Utility. *Mind, 115* (459), 607–32.

Green, M., and Hitchcock, C. (1994). Reflections on Reflection: Van Fraassen on Belief. *Synthese, 98*, 297–324.

Grice, P. (1975 (1989)). Logic and Conversation. In P. Cole and J. Morgan (eds), *Syntax and Semantics*, vol. 3 (pp. 22–40). New York: Academic Press.

Hacking, I. (1967). Slightly More Realistic Personal Probability. *Philosophy of Science, 34* (4), 311–25.

Hájek, A. (2003). What Conditional Probability Could Not Be. *Synthese, 137* (3), 273–323.

Hájek, A. (2005a). The Cable Guy Paradox. *Analysis, 65* (286), 112–19.

Hájek, A. (2005b). Scotching Dutch Books? *Philosophical Perspectives, 19* (issue on Epistemology), ed. John Hawthorne (1), 139–51.

202 REFERENCES

Hájek, A. (2008). Dutch Book Arguments. In P. Anand, P. Pattanaik, and C. Puppe (eds), *The Oxford Handbook of Rational and Social Choice* (pp. 173–195). Oxford: Oxford University Press.

Hall, N. (1994). Correcting the Guide to Objective Chance. *Mind, 103* (412), 505–17.

Hall, N. (2004). Two Mistakes about Credence and Chance. *Australasian Journal of Philosophy, 82* (1), 93–111.

Harsanyi, J. (1977). *Rational Behaviour and Bargaining Equilibrium in Games and Social Situations.* Cambridge: Cambridge University Press.

Hawthorne, J., and Lasonen-Aarnio, M. (2009). Knowledge and Objective Chance. In P. Greenough and D. Pritchard (eds), *Williamson on Knowledge* (pp. 92–108). Oxford: Oxford University Press.

Hedden, B. (2015). Time-Slice Rationality. *Mind, 124* (494), 449–91.

Hintikka, J. (1962). *Knowledge and Belief: An Introduction to the Logic of the Two Notions.* Ithaca, NY: Cornell University Press.

Hoefer, C. (1997). On Lewis's Objective Chance: Humean Supervenience Debugged. *Mind, 106*, 321–34.

Horgan, T. (2016). *Essays on Paradoxes.* Oxford: Oxford University Press.

Horgan, T. (2002). The Two-Envelope Paradox, Nonstandard Expected Utility, and the Intensionality of Probability. *Nous, 34*, 578–603.

Horton, J. (2017). Aggregation, Complaints, and Risk. *Philosophy & Public Affairs, 45*, 54–81.

Ismael, J. (2008). Raid! Dissolving the Big, Bad Bug. *Nous, 42* (2), 292–307.

Jackson, F. (1982). Epiphenomenal Qualia. *Philosophical Quarterly, 32* (127), 127–36.

Jackson, F. (2004). Why We Need A-Intensions. *Philosophical Studies, 118* (1/2), 257–77.

Jackson, F., Menzies, P., and Oppy, G. (1994). The Two-Envelope 'Paradox'. *Analysis, 54*, 43–5.

Jago, M. (2012). Constructing Worlds. *Synthese, 189*, 59–74.

Jago, M. (2014). *The Impossible: An Essay on Hyperintensionality.* Oxford: Oxford University Press.

Jeffrey, R. (1965 (1990)). *The Logic of Decision.* Chicago: University of Chicago Press.

Jeffrey, R. (1983). Bayesianism with a Human Face. In J. Earman (ed.), *Minnesota Studies in the Philosophy of Science*, vol. 10: *Testing Scientific Theories* (pp. 133–56). Minneapolis: University of Minnesota Press.

Joyce, J. (1998). A Nonpragmatic Vindication of Probabilism. *Philosophy of Science, 65* (4), 575–603.

Joyce, J. (1999). *The Foundations of Causal Decision Theory.* Cambridge: Cambridge University Press.

Joyce, J. (2007). Epistemic Deference: The Case of Chance. *Proceedings of the Aristotelian Society, 107*, 187–206.

Karni, E., and Vierø, M.-L. (2013). Reverse Bayesianism: A Choice-Based Theory of Growing Awareness. *American Economic Review, 103* (7), 2790–810.

Katz, B., and Olin, D. (2007). A Tale of Two Envelopes. *Mind, 116* (464), 903–25.

Katz, B., and Olin, D. (2010). Conditionals, Probabilities, and Utilities: More on Two Envelopes. *Mind, 119* (473), 171–83.

REFERENCES 203

Kauss, D. (Forthcoming). Context-Sensitivity and the Preface Paradox for Credence. *Synthese*.

Kelly, T. (2005). The Epistemic Significance of Disagreement. In T. Gendler and J. Hawthorne (eds), *Oxford Studies in Epistemology*, vol. 1 (pp. 167–196). Oxford: Oxford University Press.

Keynes, J. M. (1921). *A Treatise on Probability*. London: MacMillan.

King, J. C., Soames, S., and Speaks, J. (2014). *New Thinking about Propositions*. Oxford: Oxford University Press.

Kment, B. (2014). *Modality and Explanatory Reasoning*. Oxford: Oxford University Press.

Kolmogorov, A. (1933 (1950)). *Foundations of Probability*. New York: Chelsea Publishing Company.

Kripke, S. (1959). A Completeness Theorem in Modal Logic. *Journal of Symbolic Logic*, *24* (1), 1–14.

Kripke, S. (1979). A Puzzle about Belief. In A. Margalit (ed.), *Meaning and Use* (pp. 239–83). Dordrecht: Reidel.

Kripke, S. (1980). *Naming and Necessity*. Cambridge, MA: Harvard University Press.

Laplace, P. S. (1814 (1951)). *A Philosophical Essay on Probability*. New York: Dover.

Leibniz, W. G. (1666–1716 (1969)). *Philosophical Papers and Letters*. (L. Loemker, ed.) Dordrecht: D. Reidel.

Lewis, D. (1973). *Counterfactuals*. Oxford and Cambridge, MA: Blackwell and Harvard University Press.

Lewis, D. (1979). Attitudes De Dicto and De Se. *The Philosophical Review*, *88* (4), 513–43.

Lewis, D. (1986a). Introduction. In *Philosophical Papers*, vol. 2 (pp. ix–xvii). Oxford: Oxford University Press.

Lewis, D. (1986b). *On the Plurality of Worlds*. Oxford: Blackwell.

Lewis, D. (1987). A Subjectivist's Guide to Objective Chance. In D. Lewis, *Philosophical Papers*, vol. 2 (pp. 82–132). New York: Oxford University Press.

Lewis, D. (1994). Humean Supervenience Debugged. *Mind*, *103* (412), 473–90.

Lewis, D. (1996). Elusive Knowledge. *Australasian Journal of Philosophy*, *74* (4), 549–67.

Lewis, D. (1999). Why Conditionalize? In D. Lewis, *Papers in Metaphysics and Epistemology* (pp. 403–407). Cambridge: Cambridge University Press.

Liao, S.-y. (2012). What Are Centered Worlds? *The Philosophical Quarterly*, *62*, 294–316.

Mackie, P. (1987). Essence, Origin and Bare Identity. *Mind*, *96* (382), 173–201.

Mahtani, A. (2012). Diachronic Dutch Book Arguments. *The Philosophical Review*, *121* (3), 443–50.

Mahtani, A. (2016). Deference, Respect and Intensionality. *Philosophical Studies*, *174*, 1–16.

Mahtani, A. (2017). The Ex Ante Pareto Principle. *Journal of Philosophy*, *114* (6), 303–23.

Mahtani, A. (2020). Dutch Book and Accuracy Theorems. *Proceedings of the Aristotelian Society*, *120* (3), 309–27.

204 REFERENCES

Mahtani, A. (2021). Frege's Puzzle and the Ex Ante Pareto Principle. *Philosophical Studies, 178* (6), 2077–100.

Mahtani, A. (2022). The Principal Principle and the Contingent a Priori. *Manuscript.*

Meacham, C. (2010). Two Mistakes Regarding the Principal Principle. *British Journal for the Philosophy of Science, 61,* 407–31.

Meacham, C., and Weisberg, J. (2003). Clark and Shackel on the Two-Envelope Paradox. *Mind, 112* (448), 685–9.

Melia, J. (2001). Reducing Possibilities to Language. *Analysis, 61* (1), 19–29.

Mellor, D. H. (1994). *The Facts of Causation.* Cambridge: Cambridge University Press.

Nelson, M. (2002). Descriptivism Defended. *Nous, 36* (3), 408–35.

Ninan, D. (2018). Quantification and Epistemic Modality. *The Philosophical Review, 127* (4), 433–85.

Nolan, D. (2013). Impossible Worlds. *Philosophy Compass, 8* (4), 360–72.

Nolan, D. (2015). It's a Kind of Magic: Lewis, Magic and Properties. *Synthese, 197* (11), 4717–41.

Nolan, D. (2016). Chance and Necessity. *Philosophical Perspectives, 30* (1), 294–308.

Norton, J. (Forthcoming). *The Material Theory of Induction.*

Parfit, D. (1991). Equality or Priority? Lindley Lecture. University of Kansas.

Perry, J. (1977). Frege on Demonstratives. *The Philosophical Review, 86* (4), 474–97.

Perry, J. (2003). Predelli's Threatening Note: Contexts, Utterances, and Tokens in the Philosophy of Language. *Journal of Pragmatics, 35* (3), 373–87.

Pettigrew, R. (2013). Accuracy and Evidence. *Dialectica, 67* (4), 579–96.

Popper, K. (1959). The Propensity Interpretation of Probability. *British Journal for the Philosophy of Science, 10,* 25–42.

Priest, G. (2005). *Towards Non-Being: The Logic and Metaphyiscs of Intentionality.* Oxford: Oxford University Press.

Prior, A. N., and Fine, K. (1977). *Worlds, Times and Selves.* London: Duckworth.

Ramsey, F. P. (1931). Truth and Probability. In F. P. Ramsey, *The Foundations of Mathematics and Other Logical Essays* (pp. 156–98). Oxford: Routledge.

Reichenbach, H. (1949). *The Theory of Probability.* Berkeley and Los Angeles: University of California Press.

Richard, M. (1990). *Propositional Attitudes.* Cambridge: Cambridge University Press.

Ripley, D. (2012). Structures and Circumstances: Two Ways to Fine-Grain Propositions. *Synthese, 189* (1), 97–118.

Roberts, J. (2001). Undermining Underminded: Why Humean Supervenience Never Needed to Be Debugged (Even if It's a Necessary Truth). *Philosophy of Science, 68,* 98–108.

Russell, B. (1905). On Denoting. *Mind, 14* (56), 479–93.

Salmon, N. (1986). *Frege's Puzzle.* Cambridge, MA: MIT Press.

Salmon, N. (1989). Illogical Belief. *Philosophical Perspectives, 3,* 243–85.

Salmon, N. (2019). Impossible Odds. *Philosophy and Phenomenological Research, 99* (3), 644–62.

Savage, L. (1954). *The Foundations of Statistics.* New York: Wiley.

Savage, L. (1967). Difficulties in the Theory of Personal Probability. *Philosophy of Science*, *34* (4), 305–10.

Schaffer, J. (2017). Laws for Metaphysical Explanation. *Philosophical Issues*, *27* (1), 302–21.

Schervish, M. J., Seidenfeld, T., and Kadane, J. B. (2004). Stopping to Reflect. *Journal of Philosophy*, *101* (6), 315–22.

Schiffer, S. (1995). Descriptions, Indexicals, and Belief Reports: Some Dilemmas (But Not the Ones You Expect). *Mind*, *104* (413), 107–31.

Schulz, M. (2010). Chance and Actuality. *Philosophical Quarterly*, *61* (242), 105–29.

Schwartz, W. (2014). Proving the Principal Principle. In A. Wilson (ed.), *Chance and Temporal Asymmetry* (pp. 81–99). Oxford: Oxford University Press.

Schwitzgebel, E., and Denver, J. (2008). The Two Envelope Paradox and Using Variables within the Expectation Formula. *Sorites*, *20*, 135–40.

Searle, J. (1982). Proper Names and Intentionality. *Pacific Philosophical Quarterly*, *63* (3), 205–25.

Sider, T. (2002). The Ersatz Pluriverse. *Journal of Philosophy*, *99* (6), 279–315.

Skyrms, B. (2000). *Choice and Chance*. 4th edition. Ontario: Wadsworth.

Soames, S. (1987). Direct Reference, Propositional Attitudes, and Semantic Content. *Philosophical Topics*, *15* (1), 47–87.

Sobel, J. (1987). Self-Doubts and Dutch Strategies. *Australasian Journal of Philosophy*, *65* (1), 56–81.

Speaks, J. (2006). Is Mental Content Prior to Linguistic Meaning? *Noûs*, *40* (3), 428–67.

Spencer, J. (2020). No Crystal Balls. *Noûs*, *54* (1), 105–25.

Stalnaker, R. (1984). *Inquiry*. Cambridge, MA: MIT Press.

Strevens, M. (1995). A Closer Look at the 'New' Principle. *British Journal for the Philosophy of Science*, *46*, 545–61.

Sutton, P. (2010). The Epoch of Incredulity: A Response to Katz and Olin's 'A Tale of Two Envelopes'. *Mind*, *119*, 159–69.

Talbott, W. (1991). Two Principles of Bayesian Epistemology. *Philosophical Studies*, *62*, 135–50.

Teller, P. (1973). Conditionalization and Observation. *Synthese*, *26*, 218–58.

Thau, M. (1994). Undermining and Admissibility. *Mind*, *103* (412), 491–503.

Titelbaum, M. (2016). Self-Locating Credences. In A. Hájek and C. R. Hitchcock (eds), *The Oxford Handbook of Probability and Philosophy* (pp. 666–680). Oxford: Oxford University Press.

van Fraassen, B. C. (1984). Belief and the Will. *Journal of Philosophy*, *81* (5), 235–56.

van Inwagen, P. (1996). It Is Wrong, Always, Everywhere, and for Anyone, to Believe Anything, upon Insufficient Evidence. In J. Jordan and D. Howard-Snyder (eds), *Faith, Freedom, and Rationality* (pp. 137–54). Hanham, MD: Rowman and Littlefield.

Venn, J. (1876). *The Logic of Chance*. 2nd edition. London: Macmillan.

von Mises, R. (1928 (1957)). *Probability, Statistics and Truth*. English edition. London: George Allen and Unwin.

206 REFERENCES

Voorhoeve, A., and Otsuka, M. (2018). Equality versus Priority. In S. Olsaretti (ed.), *Oxford Handbook of Distributive Justice* (pp. 65–85). Oxford: Oxford University Press.

Weber, C. (2013). Centered Communication. *Philosophical Studies, 166* (S1), 205–23.

Williamson, T. (2011). Reply to Stalnaker. *Philosophy and Phenomenological Research, 82* (2), 515–23.

Yablo, S. (1993). Is Conceivability a Guide to Possibility? *Philosophy and Phenomenological Research, 53* (1), 1–42.

Yagisawa, T. (1988). Beyond Possible Worlds. *Philosophical Studies, 53*, 175–204.

Yalcin, S. (2015). Epistemic Modality De Re. *Ergo, 2*, 475–527.

Index

Note: References to footnotes are indicated by an italic "*n*" followed by the footnote number.

For the benefit of digital users, indexed terms that span two pages (e.g., 52–53) may, on occasion, appear on only one of those pages.

agency
  agent's credence function 52
  agent's epistemic state, representation
    of 48, 73–4
  choice behaviour, and 52–7
  credence claims *See* credence claims
  credence functions *See* credence
    functions
attitudes, propositional *See* propositions
attribution of beliefs *See* beliefs
attribution statements *See* credence claims

Bayes, Thomas
  'Bayesian epistemology' 1–2, 59, 74
  credence framework 47–8, 105, 137
  rationality 46–7, 74, 76
  referentialism 61
belief attribution account of metaphysically
    possible worlds 139–40, 142–6
beliefs
  attribution 51, 69, 139–40, 142–6
  conflicting 63
  credences, and 1, 190*n*30
  desires, and 29–30, 56–7
  mediation of 51
  objects, about 51, 54–5
  propositional attitudes, as 29–30
  propositions, and 12, 29–30
  true or false 12
  truth-values 54–5
  unexpressed 8–9
  utterances, and 12
better restrictions, decision theory
    and 118–23
betting behaviour, choice behaviour and 36,
    47–8

card games, as illustration of Reflection
    Principle 77–9
Chalmers, David 60, 139–40, 146–52, 154,
    162, 164–8, 193–4
chance claims
  credence framework, and 93
  credences, and 87, 104
  hyperintensionality of 93, 104
  intensionality of 93, 104
  non-extensionality of 88–92
  opacity of 87
chance, credences and 87
chance framework 92–5, 102–3
chance function
  conditionalisation of 100–3
  credence function, and 97
  deference to 46, 73, 87, 94–5, 97,
    100–1
  metaphysically possible worlds,
    and 92–3
  New New Principle, and 102
  non-hyperintensionality of 96
  Principal Principle, and 94–5
  tracking of 100
choice behaviour
  actual and dispositional 36–7
  betting behaviour, and 36, 47–8
  credence claims and 52–7
  credence function, and 52
  credences, and 36–7, 68*n*15
  epistemic states, and 40
  opacity of credence claims, and 52, 72
coherence notion as to states 188–91
coherence requirement for states 184–8
completeness requirement for
    states 184–8

208 INDEX

conditionalisation
  chance function, of 100, 102–3
  credence claims 60–2
  definition of 73
contingent a priori, Principal Principle
    and 94–7
credence claims
  author's conclusions summarized 72
  choice behaviour, and 52–7
  conditionalization 60–2
  decision theory *See* decision theory
  guise-based account of credence 66–9
  guise-based account of credence,
    alternative version 69–72
  Guise Russellianism 62–6
  'it's just obvious' argument 49–52
  'logical omniscience' problem 57–60
  opacity of 49, 73–4, 131–5, 137–8, 169,
    196–7
  practical implications of opacity 105
  referentialism 60–1
credence framework
  agent's epistemic state, representation
    of 48
  author's approach and argument
    summarized 4–6
  author's conclusions summarized 197–8
  awareness of 4–5
  Bayesian framework 47–8
  claims of 2, 4
  content and structure of current
    book 3–4
  credences, nature of 35–7
  descriptive outline of 3–4, 37–40
  diachronic rules 44
  encounter with 1–2
  example illustration of 2–3
  flexibility of 5
  foundations of 4
  intended readership of current book 5–6
  interpretation of 4, Chapter 7, Chapters
    7 & 8
  introduction to 1, 3–4, 31–2
  'logical omniscience' problem 57–60
  objects of credence, fine-grained nature
    of 74
  opacity of 2, 4–5
  opacity of credence claims 49
  philosophy of language, and 2
  probability axioms 40–4

  probability framework 32–5
  propositions, and 30
  rationality rules relevant to 44–7
  reassessment of 5
  synchronic rules 44
  transparency of 2
  users of 4–5
  users' responses to 5
  uses of 3–5
  viable interpretation of 4
credence function
  agency 52
  betting behaviour, and 52–4
  choice behaviour, and 52
  Clarified Generalized Reflection Principle,
    and 81–2
  conditionalisation of 44–7, 73, 99–100,
    102–3
  constraint of 44–6
  deference principles 99–100
  domain of 163
  guise-based account of credence,
    and 67–72
  hyperintensionality of 96
  metaphysically possible worlds, and 138–9
  MEU, and 108
  objects of 37–8, 67–9
  objects of credence, and 153, 165, 197–8
  Principal Principle, and 92–3, 97,
    104*n*21
  probabilism, and 192–3
  probability axioms, and 40–4, 46–7, 57–8,
    73–4, 152, 185–6
  probability function as 37–9
  representation of 38
  Restricted Reflection Principle, and 77
  sentence-context-worlds, and 178–9
  synchronic rules, and 44
  two-envelope paradox, and 120–1
  welfare economics, and 125–6
credences
  beliefs, and 190*n*30
  chance, and 87
  nature of 35–7

decision theory 4
  author's conclusions summarized 136–7
  better restrictions 118–23
  decision tables 117–18
  definite descriptions 115–17

INDEX 209

egalitarianism, and 128
illustrative example of 105–8
'maximize expected utility' rule
    (MEU) 107–8, 111, 125–37
objects of credence, and 108–11
opacity of credence claims, and 131–5
outcomes for groups of people 125–7
outcomes for individuals 131–5
Pareto principle, and 128–31
prioritarianism, and 128
rigid designators 117–18
supervaluationism, and 135–6
transparent designators 114–15
two-envelope paradox 4, 111–25
two-envelope paradox, specific version
    of 113–14
utilitarianism, and 127–8
variations 123–5
welfare economics, and 4, 125–36
declarative utterances, propositions as
    contents of 10–12
deference principles 86–7
definite descriptions 115–17
descriptivist account of metaphysically
    possible worlds 139–42, 168, 182–3
desires
beliefs, and 56–7
propositions, and 29–30
diachronic rules 44
disagreement principles 86–7

economics See welfare economics
egalitarianism, decision theory and 128

fairytales as illustration of Reflection
    Principle 84–6
fine-grained worlds 170–5
Frege, Gottlob 13–18, 20–2, 24, 26–7, 28n16,
    30, 49–52, 62, 74n2, 198

Guise Russellianism
application of 59n8
credence claims 62–6
Frege's account, and 27–8
guise-based account of credence 66–9
guise-based account of credence,
    alternative version 69–72
propositions, account of 20–2, 30

'it's just obvious' argument 49–52

language
credence framework, and 2
familiarity with 4–5
linguistic representations 175–7
propositions, and 30
propositions, nature of 3–4
sentence-context-worlds 179–81
sentence-worlds 177–9
world-building languages 181–4
See also language (philosophy of)
linguistic representations 175–7
'logical omniscience' problem 57–60

'maximize expected utility' rule
    (MEU) 107–8, 111, 125–37
metaphysically possible worlds
author's approach and argument
    summarized 139–40
author's conclusions summarized 168
Chalmers's two-dimensionalism account,
    and 139–40, 146–52, 154, 162, 164–8
credence claims, new convention 157–62
introduction to 138–40
objects of credence as enriched
    propositions 166–8
objects of credence as primary
    intensions 152–7, 162–6
objects of credence as secondary
    intensions 162–6
Russell's descriptivism, and 139–42
Stalnaker's belief attribution account,
    and 139–40, 142–6
states as 138
states as not 169–70, 196

objects, beliefs about 51, 54–5
objects of credence
decision theory, and 108–11
enriched propositions, as 166–8
fine-grained nature 74
primary intensions, as 152–7,
    162–6
secondary intensions, as 162–6
opacity of credence claims See credence
    claims

Pareto principle 128–31
philosophy of language See language
playing cards, dealing of, as illustration of
    Reflection Principle 77–9

210 INDEX

possible worlds (semantics of)
  propositions 22–9
  See also metaphysically possible worlds
Principal Principle
  author's approach and argument
    summarised 88
  author's conclusions summarized 104
  chance claims, non-extensionality
    of 88–92
  chance framework 92–4
  connection between credences and
    chance 87
  contingent a priori, and 94–7
  New New* Principle 103
  New New Principle 101–4
  New Principle 100–1
  old problems for 97–101
  opacity of credence claims, and 87, 104
  Reflection Principle, and 87
principles
  deference principles 4, 86–7
  disagreement principles 86–7
  Principal Principle *See* Principal
    Principle
  Reflection Principle *See* Reflection
    Principle
prioritarianism, decision theory and 128
probabilism
  definition of 73
  probabilism-based arguments as to
    states 191–3
  rationality, and 73–4
probability axioms
  credence framework 40–4
  credence functions, and 73–4
  rationality, and 73–4
probability framework, credence framework
  and 32–5
propositions
  beliefs, and 29–30
  controversies as to nature of 30
  credence framework, and 30
  declarative utterances 10–12
  desires, and 29–30
  Frege's account of 13–18, 30
  Guise Russellian account of 20–2, 30
  introduction to 7
  nature of 3–4
  philosophy of language, and 30
  possible worlds (semantics of) 22–9

propositional attitudes 7–10, 29–30
  roles of 7–13
  Russell's account 18–20, 30
  structured propositions 193–6
  truth-aptness of 12–13
  See also beliefs; desires

rationality
  author's approach and argument
    summarized 74
  conditionalization *See* conditionalization
  credence claims *See* credence claims
  credence functions *See* credence
    functions
  introduction to 73–4
  principles of *See* principles
  probabilism 73–4
  rules of, credence framework
    and 44–7
referentialism, credence claims and 60–1
Reflection Principle
  Clarified Generalised Reflection
    Principle 81
  controversy as to 74–5
  dealing of playing cards as illustration
    of 77–9
  definition of 73
  Generalised Reflection Principle,
    and 75–7
  Improved Generalised Reflection
    Principle 79–84
  Restricted Reflection Principle 77
  'Sleeping Beauty' story as illustration
    of 84–6
rigid designators 117–18
Russell, Bertrand 18–20, 27–8, 30, 139–42,
  168, 182–3
  See also Guise Russellianism

sentence-context-worlds 179–81
sentence-worlds 177–9
'Sleeping Beauty' story as illustration of
  Reflection Principle 84–6
Stalnaker, Robert 139–40, 142–6
states
  coherence notion 188–91
  coherence requirement 184–8
  completeness requirement 184–8
  fine-grained worlds 170–5
  linguistic representations 175–7

INDEX 211

metaphysically possible worlds, as
  See metaphysically possible worlds
metaphysically possible worlds, as
  not 169–70, 196
probabilism-based arguments 191–3
sentence-context-worlds 179–81
sentence-worlds 177–9
structured propositions 193–6
world-building languages 181–4
structured propositions *See* propositions
supervaluationism 135–6
synchronic rules 44

transparent designators 114–15
truth
  truth-aptness of propositions 12–13
  truth-values, beliefs and 54–5

two-dimensionalism account of
  metaphysically possible worlds 139–40,
  146–52, 154, 162, 164–8
two-envelope paradox Chapter 6
  decision theory and 113–14
  problem of 4

uncertainty, experience and 1
utilitarianism
  decision theory and 127–8
  'maximize expected utility' rule
  (MEU) 107–8, 111, 125–37,

welfare economics
  decision theory, and 4
  decision theory and 125–36
world-building languages 181–4